# Vehicular-2-X Communication

Radu Popescu-Zeletin · Ilja Radusch ·
Mihai Adrian Rigani

# Vehicular-2-X Communication

State-of-the-Art and Research in
Mobile Vehicular Ad hoc Networks

 Springer

Radu Popescu-Zeletin
Fraunhofer-Institut für Offene
Kommunikationssysteme
(FOKUS)
Kaiserin-Augusta-Allee 31
10589 Berlin
Germany
radu.popescu-zeletin@fokus.fraunhofer.de

Ilja Radusch
Fraunhofer-Institut für Offene
Kommunikationssysteme
(FOKUS)
Kaiserin-Augusta-Allee 31
10589 Berlin
Germany
ilja.radusch@fokus.fraunhofer.de

Mihai Adrian Rigani
Fraunhofer-Institut für Offene
Kommunikationssysteme
(FOKUS)
Kaiserin-Augusta-Allee 31
10589 Berlin
Germany
mihai.adrian.rigani@fokus.fraunhofer.de

ISBN 978-3-642-42605-6          ISBN 978-3-540-77143-2  (eBook)
DOI 10.1007/978-3-540-77143-2
Springer Heidelberg Dordrecht London New York



*Cover design*: eStudio Calamar S.L.

Printed on acid-free paper

Springer is part of Springer Science+Business Media (www.springer.com)

# Contents

# Chapter 1
# Introduction

Universal vehicular communication promises many improvements in terms of accident avoidance and mitigation, better utilization of roads and resources such as time and fuel, and new opportunities for infotainment applications. However, before widespread acceptance, vehicular communication must meet challenges comparable to the trouble and disbelief that accompanied the introduction of traffic lights back then. The first traffic light was installed in 1868 in London to signal railway, but only later, in 1912, was invented the first red-green electric traffic light. And roughly 50 years after the first traffic light, in 1920, the first four-way traffic signal comparable to our today's traffic lights was introduced.

The introduction of traffic signals was necessary after automobiles soon became prevalent once the first car in history, actually a wooden motorcycle, was constructed in 1885. Soon, the scene became complicated, requiring the introduction of the "right-of-way" philosophy and later on the very first traffic light.

In the same way the traffic light was a necessary mean to regulate the beginning of the automotive life and to protect drivers, passengers, as well as pedestrians and other inhabitants of the road infrastructure, vehicular communication is necessary to accommodate the further growth of traffic volume and to significantly reduce the number of accidents.

Vehicular communication cannot only create an extended virtual information horizon, warning drivers early of dangers ahead and thereby avoiding accidents, but also allows for mitigation of unavoidable accidents by advanced short-range communication between the cars involved. Furthermore, as the systems stabilize, vehicle communication can evolve into paradigms such as cooperative driving analog to traffic lights evolved from man-controlled lamps to automatic traffic management systems. Already, today's literature suggests that cooperative driving is more efficient, provides more safety and improvements of traffic-flow stability. Cooperative behavior is very beneficial to improve existing applications such as automatic adaptation of speed with the vehicle in front (each vehicle is driven by a human driver) or automatic following of a leading vehicle forming a platoon of vehicles (only leader is driven by human driver). But cooperative behavior can go much further, providing for the first time a possible solution to achieve applications such as collision avoidance or automatic merging of vehicles on the highway, which without vehicular communication were only a dream. As of today, the goal of cooperative driving is

R. Popescu-Zeletin et al., *Vehicular-2-X Communication*,
DOI 10.1007/978-3-540-77143-2_1, © Springer-Verlag Berlin Heidelberg 2010

to correct errors or deviations from the required behavior in dangerous situations on behalf of the driver.

## 1.1 Overview

This book describes the various aspects of vehicular communication such as medium access control, routing, security, and accompanying standards along the ISO OSI reference model. Furthermore, future automotive applications such as cooperative driving maneuvers utilizing vehicular communication are introduced and described in detail. Moreover, orthogonal to this description of vehicular communication technologies, a novel testing and simulation approach combining current approaches for traffic and network simulations is introduced as a method for validating the introduced automotive applications. Research and development projects are also outlined. We strongly believe that the results will be introduced soon due to:

- Traffic flung
- Costs (humanly and materially) of the accidents
- Stronger need of environment preservation

Traffic flung refers to increased and still increasing traffic in all industrialized countries all over the world, which leads to the increasing number of accidents. Therefore, one main objective of the European Union is to reduce traffic accidents by half in the end of 2010.

In order to maximize the safety benefits gained from new vehicle technology, the focus should be both on innovation and implementation.

## 1.2 Why Vehicular Communication?

In the beginning of vehicle industry, streets and different types of vehicles were considered as autonomous systems. Later, the influence of each other only in size, but sharing common resources, introduced regulations on the streets that, again, were considered autonomous vehicle systems. Today, the number of accidents and traffic jams is increasing, due to a continuous growing number of vehicles and an imperfect resource sharing. The next step would be to create a new way of controlling the system. This refers to new kind of applications that will enable the fully automated car of the future to drive by itself. Today, the focus lies on the automation of specific difficult driver maneuvers. These maneuvers, without automation, are leading to a growing number of accidents. Examples of such applications are: automatic speed adaptation to follow a leader on the highway, automatic entrance on the highway or self parking. To achieve such tasks, a great step forward would be to use environmental perception of surroundings by vehicle communication in the so called cooperative approach [1, 2].

Research on intelligent transport systems dates back in the late 1980s and beginning of 1990s. Since its beginning, research shifted from Automated Highway Systems (AHS) to the Intelligent Vehicle Initiative (IVI), trying to improve Safety (drivers and other road users), Resource Efficiency (use of roads as well as use of fuel) and Infotainment/Advanced Driver Assistant Systems (ADAS).

Since the beginning of research in intelligent transport systems, there have been three areas that can possibly be improved: Vehicles (e.g. adaptive cruise control, collision avoidance systems), Roads (e.g. adaptive speed control, advanced traffic management), and Drivers (e.g. by providing advance traffic information, collision warnings). Of course, most research projects followed a mixed approach, trying to improve all three areas.

Furthermore, the general feasibility and technological maturity of adaptive cruise control is also demonstrated by vehicle manufacturers as they already include features like detecting the leading vehicle and maintaining appropriate distances as well as support for stop-and-go traffic in their luxury-lines of cars. However, the long term impact of these technologies to road safety and traffic efficiency has yet to be verified in, e.g., large scale field-tests. Again, current research investigates how inter-vehicle communication can increase the proactive safety and is outlined in more detail below.

Examples of the vehicular communication benefits are applications such as different warnings (on road incidents or traffic alerts) as well as improvement of classic applications such as automatic adaptation of speed according to the vehicle in front (Cooperative Adaptive Cruise Control), assistant merging of vehicles on the highway (Cooperative Merging), assistant follow of a leading vehicle (Cooperative Platooning), assistant avoidance of collisions (Cooperative Collision Avoidance). The word "Cooperative" from the above applications means that vehicles cooperate with each other by exchanging information by means of vehicle to vehicle communication. This communication offers features such as: full 360° around vehicle situation awareness that is far more reliable than local sensors and provides a bigger coverage area in all directions, warnings about different hidden hazards such as accidents or obstacles behind a curve. Such features are unbeatable compared to existent technologies such as normal radar based sensors. One big drawback of the sensors is that they are influenced by weather conditions and by mud or dirt.

The vehicular communication opens up new features, but let's see how this technology can be implemented. In the next chapter, we present a fundamental design of such a system.

## 1.3 Architecture Layers

The transmission of the information in the vehicular network may be done in multiple ways according to the application. We define three communication regimes: bidirectional, single-hop and multi-hop position based. Each regime is designed for a specific communication. Bidirectional is a classic communication in both directions. The other two regimes are one-way communications. Single-hop is a fast one,

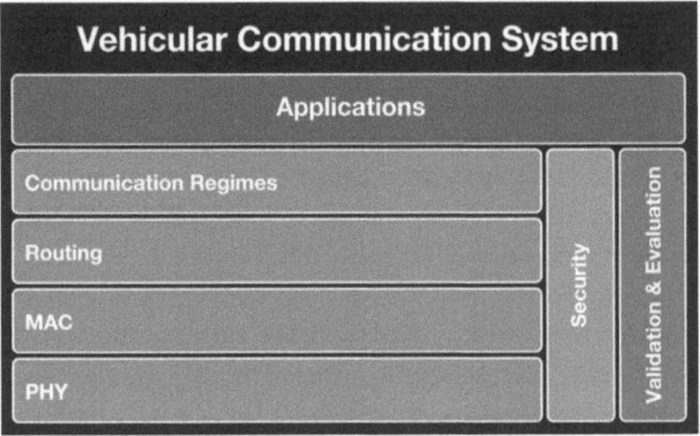


**Fig. 1.1** Communication layers architecture

while multi-hop is a slower one. All three regimes are detailed in Sects. 3.1, 3.2, and 3.3.

In order to send all the vehicular information, used in applications, we need a protocol that handles the messages transmitted and tries to avoid message collisions. This is called the MAC (Medium Access Control) layer.

To transmit messages in the vehicular network, we also need physical channel(s) with dedicated frequency range(s). This is called the physical layer (abbreviated PHY). We present here the future standard to be used in vehicular communication, called Dedicated Short Range Communications (802.11p).

To move a data packet from source to the destination, we need a routing layer.

We argue that all of the communication regimes require reliable security that is included in a security layer.

An architecture is the "fundamental organization of a system embodied in its components, their relationships to each other, and to the environment, and the principles guiding its design and evolution" [3].

In Fig. 1.1 we present an overview of the vehicular communication architecture.

On top, we put the applications that arise due to the new communication system. Then, all the layers in the communication system from high level down are presented. On top we have the communication regimes. Then we have the rest of the layers: Routing, MAC (Medium Access Control) and PHY (Physical). Orthogonal to these is the security layer.

At first, we will begin to present the applications in the vehicular communication.

# References


1. L. Andreone and M. Provera, Inter-vehicle communication and cooperative systems: local dynamic safety information distributed among the infrastructure and the vehicles as "virtual sensors" to enhance road safety, http://www.car-to-car.org
2. D. van Arem, C.J.G. van Driel, and R. Visser, The impact of cooperative adaptive cruise control on traffic-flow characteristics, 2006
3. IEEE, IEEE standard 1471-2000: IEEE recommended practice for architectural description of software-intensive systems, 2000

# Chapter 2
# Applications of Vehicular Communication

Summarizing recent classifications (such as the Car-2-Car Communication Consortium), we will, for the purpose of this book, use the following extended top-level application domains (Fig. 2.1):

- Safety,
- Resource Efficiency (including traffic as well as environmental efficiency), and
- Infotainment and Advanced Driver Assistance Services (ADAS).

Safety domain refers to applications or systems that increase the protection of the people in the vehicle as well as the vehicle itself. The systems shall save lives by avoiding or minimizing the effects of an accident.

Resource Efficiency refers to increase traffic fluency and is of great interest today as much as in the future, since traffic congestion is becoming an increasingly severe problem. Better traffic efficiency results in less congestion and lower fuel consumption, helping to minimize environmental and economic impact.

Infotainment and Advanced Driver Assistance Services provide entertainment or information to drivers and passengers. This makes driving more comfortable by providing access to different services such as easy toll payment without stopping. Entertainment examples are the possibility of playing music, making phone calls, or listening to text messages by using wheel-mounted buttons, touch-screen interface or through voice commands. Infotainment and ADAS also refers to currently implemented systems such as navigation systems or hands-free systems as well as providing a smart interface to connect vehicles with cell phones, PDAs or iPods.

Former approaches for classification of car related research use cases include the one proposed by the Automotive Multimedia Interface Collaboration (AMI-C) in 2003 [1]. Their classification has later been adopted by OSGi, ISO TC222, Bluetooth, ITU-T as well as the 1394 Trade Association. AMI-C defined 18 top level categories (plus one additional combined category). The communication technology proposed for use by AMI-C ranges from off-the-shelf radio/TV tuners to mobile phones to manufacturer provided V2V communication hardware.

R. Popescu-Zeletin et al., *Vehicular-2-X Communication*,
DOI 10.1007/978-3-540-77143-2_2, © Springer-Verlag Berlin Heidelberg 2010

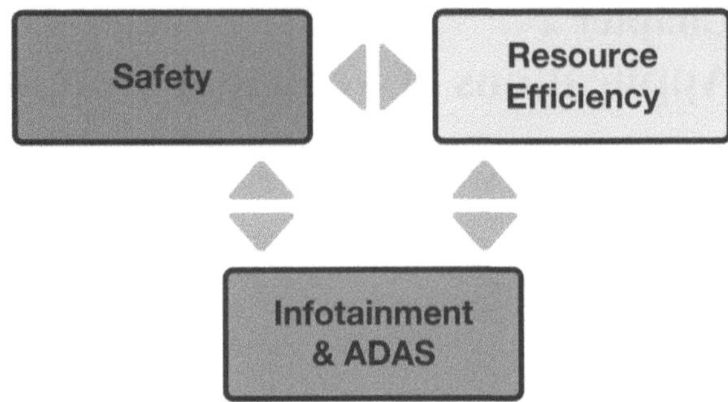


**Fig. 2.1** Application domains

## 2.1 Safety

We are defining the most frequent traffic situations leading to accidents by analyzing accident statistics. We point out that a system to avoid an accident based on vehicular communication offers unforeseen benefits in terms of other vehicle detection and avoidance systems.

Considering today's vehicle accident statistics worldwide, it can be concluded that in 2004 about 1 million people were killed and about 40 million were injured [2]. Efforts are made to lower the number of accidents, but with the continuously increasing number of vehicles and with today's technology this is very difficult to achieve. Below, we are presenting an overview of different types of collisions that may occur: [3–5]:

- Head-on collisions (with other vehicles)
- Rear-end collisions (with other vehicles)
- Side collisions (with other vehicles)
- Vehicle collision with fixed objects (e.g.: with a tree)
- Vehicle collision with bicycles
- Vehicle collision with pedestrians
- Vehicle collision with animals
- Rollovers (e.g.: inadequate speed in curve)
- Level crossing accidents (railroad crossing)
- Multi-vehicle collisions

Road fatalities in 2004 in the European Union (EU) (excluding Germany) depending on transportation type as identified in [6] are presented in Fig. 2.2. It can be seen that car fatalities are leading (54%), followed by pedestrian (14%), motorcycle/moped (20%), bicycle (5%) and lorries (5%).

It is interesting to note, that more than 80% of the fatalities occur despite good weather conditions (Fig. 2.3) [6]:

The life of the persons in a vehicle depends on the driver's reaction to sudden events, such as the appearance of a traffic jam end. The chances of an accident worsen if other factors, such as bad weather conditions (e.g. fog, rain) or lack of

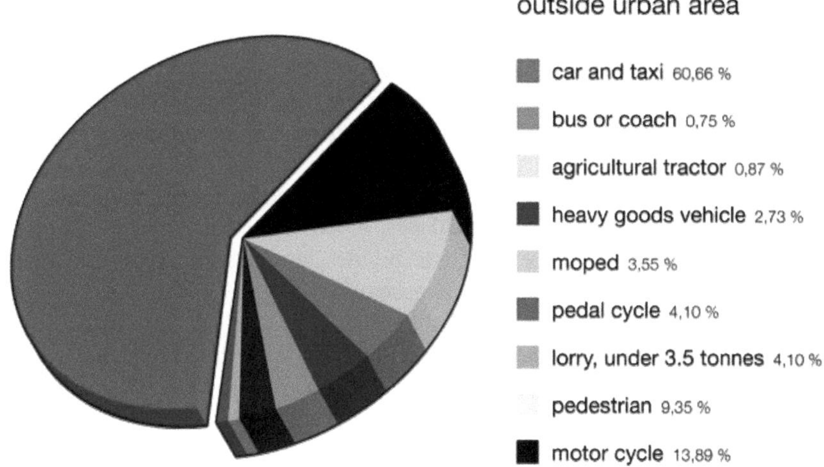


**Fig. 2.2**  Total accidents in EU by transportation type [6]

**Fig. 2.3**  Total accidents in
EU by weather conditions in
2004 [6]

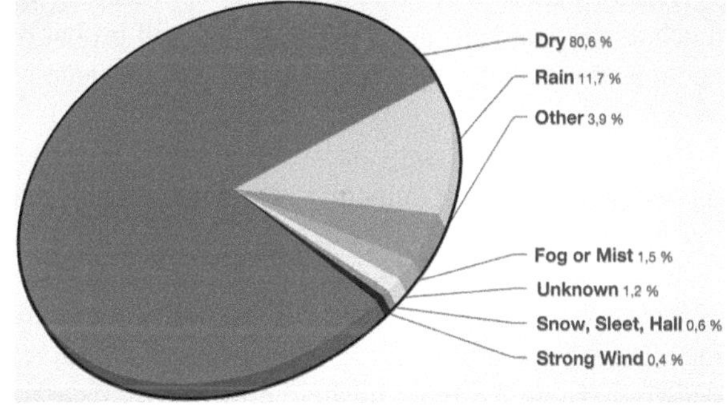


visibility (e.g. due to a curve) occur. Even more, human drivers can make mistakes
or react too slowly due to different subjective reasons (bad visibility, inexperience,
tiredness or drunkenness). But the reaction time of an automated system cannot be
beaten even by the most experienced driver. Take, for example, the common failure
to push strong enough on the brake pedal in emergency situations. Therefore, the
half-autonomous brake assistant feature was introduced to enable a full power brake
based on the speed with which the pedal is pressed.

Safety on the road has always been an issue, but today due to the continuously
growing number of vehicles, the number of accidents also increases and the need for
better safety systems becomes essential. It is unfortunate that the automobile, which
transformed the modern world by offering mobility and autonomy, led to such an
enormous loss in human lives. But this can be changed.

Today, research and development work [7] has been done to face road safety
problems by developing driver assistance systems, based on sensor technologies

that are able to detect the traffic situation surrounding the vehicle and, in case of danger, to warn the driver. These systems are called active safety applications. It is a start, but relying only on local sensors is not enough.

The solution is to raise active safety technology to the next level, which we are calling proactive safety, enabling vehicles to communicate and coordinate responses to avoid collisions. Our goal is that accidents become as rare as plane crashes are today. The European Commission expresses in their safety program the vision to have "zero road fatalities" by 2020 [8]. In order for this "vehicle surrounding situation awareness" to be effective, besides inter-vehicle communication, a vehicle-infrastructure communication is necessary.

The roadside infrastructure, e.g. traffic lights, road signs, etc., needs to be equipped with communication counterparts [9]. This requires either direct investment by the car manufacturers or by the government agencies. The features of vehicle-to-roadside can go further, like transmitting alerts to traffic lights about vehicle location and speed, forcing the lights to adapt to traffic demands. For instance, during rush hours, cars may inform a traffic light to alter the intervals between light changes. If required, the traffic light would shorten the time the light is red and prolong the time it is green to process cars more quickly.

The introduction of proactive safety applications will be needed in the future as much as the seat belts are today. But this will probably take a long time. The introduction of seat belts took decades and only became effective once it was required by law.

But cooperative safety applications need a certain penetration rate to work. Showing the impact of vehicular communication in safety domain applications, this rate represents the percentage of equipped vehicles and it is depicted in Table 2.1 [9].

It is seen that warnings can be accomplished only with a low penetration rate of above 10%, while cooperative driving requires too much (100%) penetration rate. The figures are based on simulations and field tests and represent the starting penetration rate to create positive effects, i.e.: have more time to avoid a possible collision. Due to the fact that cooperative driving takes off the responsibility from the driver for certain actions, it has to be 100% reliable. This implies a requirement of 100% penetration rate.

Based on the required penetration rate, first we divide the applications in cooperative behavior and warnings and present a short overview of what cooperative actually means. A detailed classification of safety domain is presented in Sect. 2.1.2.

Cooperative behavior refers to the ability of vehicles to cooperate with each other by means of communication, so that the intentions and positions of other vehicles are known to each other. Vehicles need to share the information they posses.

**Table 2.1** Penetration rates of applications

| Applications | Required penetration rate |
| --- | --- |
| Local danger warning | Low (above 10%) |
| Co-operative driving | High (100%) |

Cooperative behavior enhances the perception of environments not only through its own sensors, but through the sensors of other vehicles. Some examples of cooperative applications are: Cooperative Merging, Cooperative Adaptive Cruise Control or Cooperative Forward Collision Warning (depicted in Sect. 2.1.2).

Different warnings refer to on-road incidents or traffic alerts in order to improve traffic fluency or even safety by avoiding accidents. Some applications are warnings of hazards, approaching emergency vehicles, slow vehicles, or traffic jams. Of course, a reliable dissemination of all this alerts can be accomplished by vehicular communication.

The applications of vehicle communication presented above offer great benefits to the drivers. Summarizing, they create new information or share existing information in a way that was not feasible before.

## 2.1.1 Critical Traffic Situations

Below you will find different traffic situations of potential accidents to be avoided. These situations can be classified based on (Fig. 2.4):

- type of road/place (straight, curved, intersection, straight road)
- type of "obstacle" (vehicles, motorcycles, bicycles, pedestrians, animals, trees, pillars)
- relative position of the sides involved (narrow distance between vehicles, head-on, rear-end, side)

The third classification is depicted in detail below, as vehicular communication brings enhancements in these scenarios. In all of the relative positions of the sides involved, the following detection systems are required:

- v2v communication,
- radar based,
- camera vision,
- GPS with digital map.

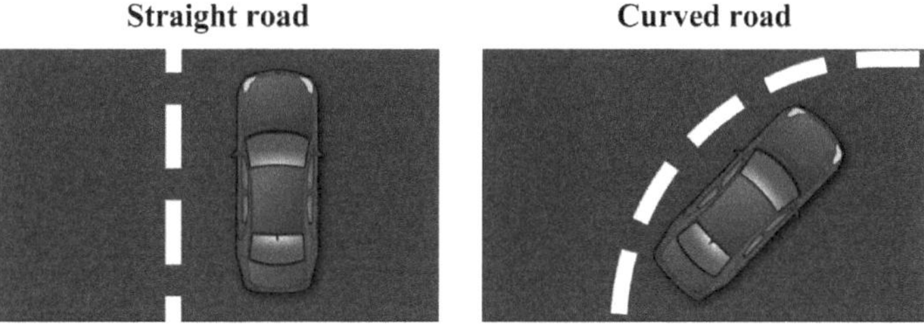


**Fig. 2.4** Classification of road situations

### 2.1.1.1  Cars Passing at Close Distance (Head-on)

This scenario must take into account special needs such as minimization of false alarms to avoid misinterpretation of this scenario with head-to-head collision. On the other hand, lateral "scratches" that may occur must be avoided. For all this to be achieved, a real-time update of all movement data parameters (position, velocity, heading, etc.) must be kept. A typical scenario: two vehicles are "forced" to pass close each other on a narrow bridge (Fig. 2.5).

### 2.1.1.2  Head-on Vehicle Collision

This situation can typically be found in an overtaking maneuver. These are the most dangerous collisions as the relative speed between vehicles is extremely high – actually, the sum of both vehicles speeds (Fig. 2.6).

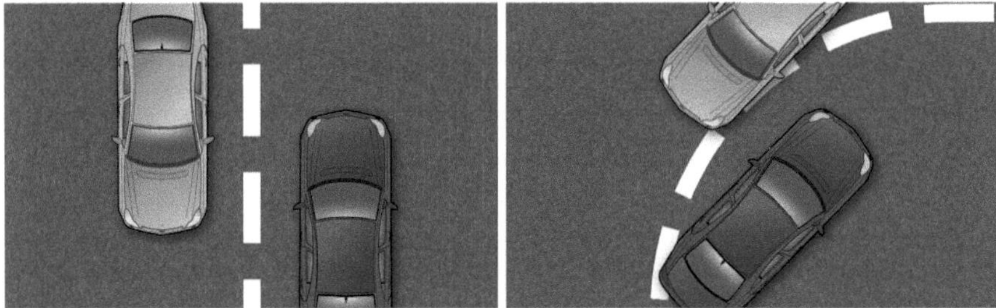

**Fig. 2.5**  Passing cars

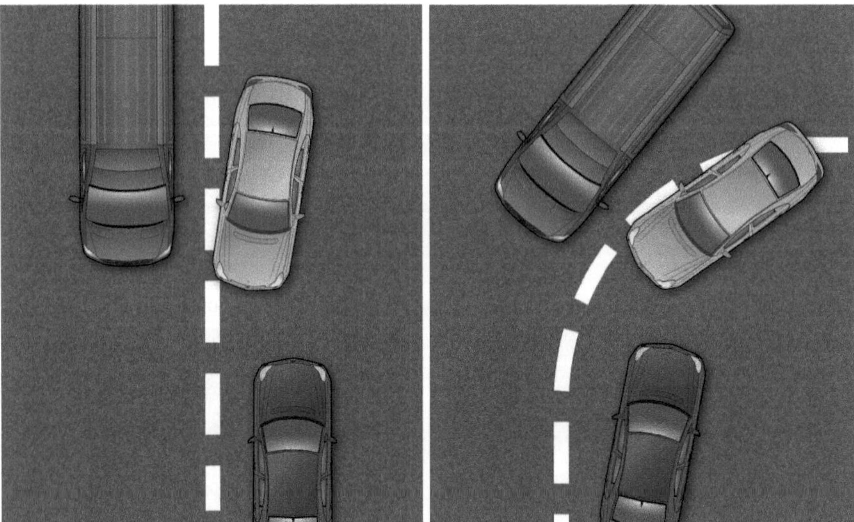

**Fig. 2.6**  Head-on vehicle collision

At the time the two vehicles that meet head to head reach the "last point of stopping", the following should be analyzed:

- If the overtaking maneuver can be successful completed (based on parameters of both vehicles: acceleration, speed – see above) => no interference
- If it cannot be successfully completed, then other solutions are taken into account (in this order):

  - Brake and make the appropriate maneuver just to re-enter in an "empty" space in the back of the overtaken car,
  - Full stop, maybe even reverse if time allows it (if the previous action cannot be completed). This would be an extreme solution just to minimize the impact.

For two vehicles in head-on collision danger that have each a speed of 120 km/h the stopping distance between them is about 7 s (140 m) on dry asphalt.

The distance of an overtaking maneuver depends on the acceleration (car model, speed) but may take from 100 m to a couple of hundred meters, as identified in Table 3.4 and [10].

### 2.1.1.3 Rear-End Vehicle Collision

This type of collision accounts for more than 17% of accidents involving fatalities or injuries in Germany [11]. According to [6] in 1998 in EU (10 countries, Germany not included) 13.3% of all accidents (injured and killed) are rear end accidents. Also, this type of collision is more often seen on highways than on rural and urban roads (Fig. 2.7).

"The last point of stopping" should be set for each car model and current speed accordingly. This way the stopping distance is always kept. So, for example, if a vehicle is driving at 120 km/h, it will know its stopping distance is 90 m, while another newer vehicle has a stopping distance of only 60 m. One problem occurs on roads with one lane on each side, where overtaking may be required and by keeping always a distance of e.g. 90 m to the car in front would prolong the overtaking maneuver. But, what will happen if not all cars have a collision avoidance system? Then the front car should also monitor its back to determine if the rear vehicle is unable to stop in time (v2v communication).

Generally, a radar-based sensor (with a range of 150 m) [12, 13] may be sufficient to stop a vehicle from a speed of 160 km/h (if no human delays are added).

If "the last point of stopping" is passed deliberately (e.g. to shorten an overtaking) and the rear vehicle is unable to fully stop in time, the following should be analyzed:

- brake and make the appropriate maneuver just to avoid the vehicle (lateral radar must be engaged to be sure it can make the maneuver). The best solution has to be determined, e.g. evasive maneuver is engaged if on its lane will hit the front

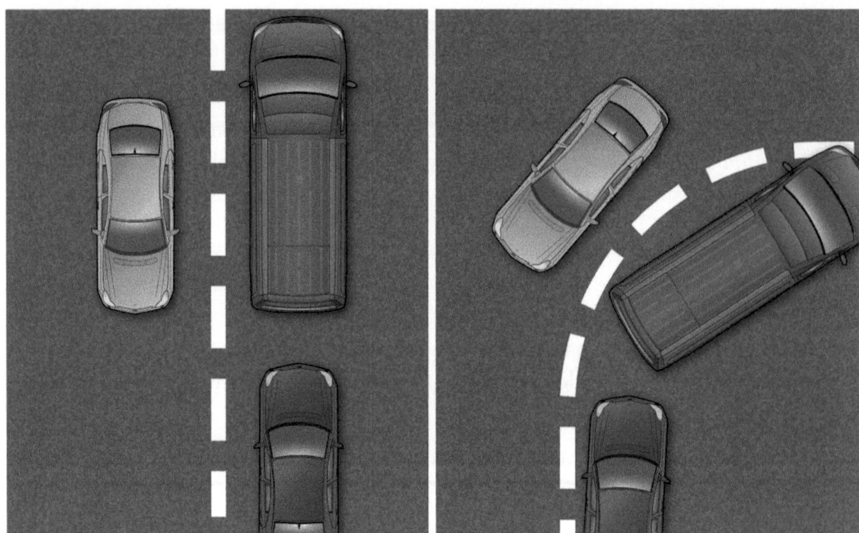

**Fig. 2.7**  Rear-end vehicle collision

vehicle by 80 km/h and on a side lane would only hit it by 50 km/h (the speeds of both vehicles in collision are taken into account).

### 2.1.1.4  Side Collisions at Intersections

Different types of intersections exist. The side collision may not necessarily be on a perpendicular angle. Disregarding any sign in intersection, an avoidance of collision only based on sensors is generally impossible. In order to avoid collision in intersections, communication with infrastructure (stoplight) and other cars is required. Visual recognition of traffic signs may also be implemented. The collision avoidance based only on sensors may work on the principle from "rear-end" scenario, but will only be successful if one of the vehicles is already in an intersection while the other one is approaching the intersection and has enough time to fully stop (Figs. 2.8, 2.9, 2.10, 2.11, 2.12, 2.13, and 2.14).

The "merge lane" is a special scenario of an intersection. To allow vehicles to join ("on-ramp") a flowing traffic with no collision. This has two "sub-scenarios":

- a "lane change" collision. If the vehicles are driving in parallel and one has to enter (the on-ramp lane finishes) then lateral sensors are required with priority on the blind spots (see Fig. 2.15). Based on these sensors only, it is hard to make a decision. Therefore, to create a reliable collision avoidance system in this situation, a v2v communication system is beneficial (in order to establish a priority / agreement between vehicles).
- if the vehicle is already entered in front, then this scenario becomes a rear-end collision avoidance as described in Sect. 2.1.1.3.

**Fig. 2.8**  Perpendicular intersection

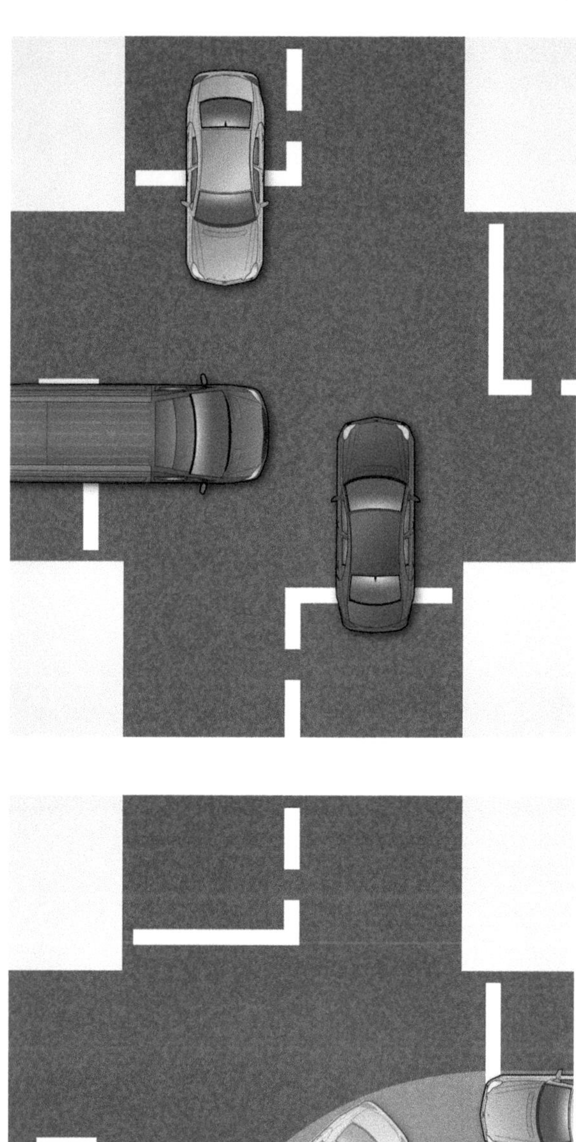

**Fig. 2.9**  Left turn

**Fig. 2.10** Left turn with two lanes on one way

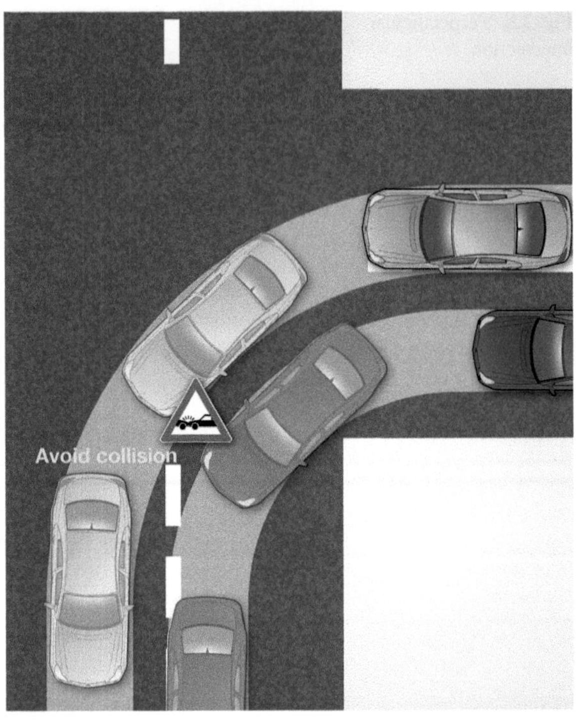


**Fig. 2.11** Right turn

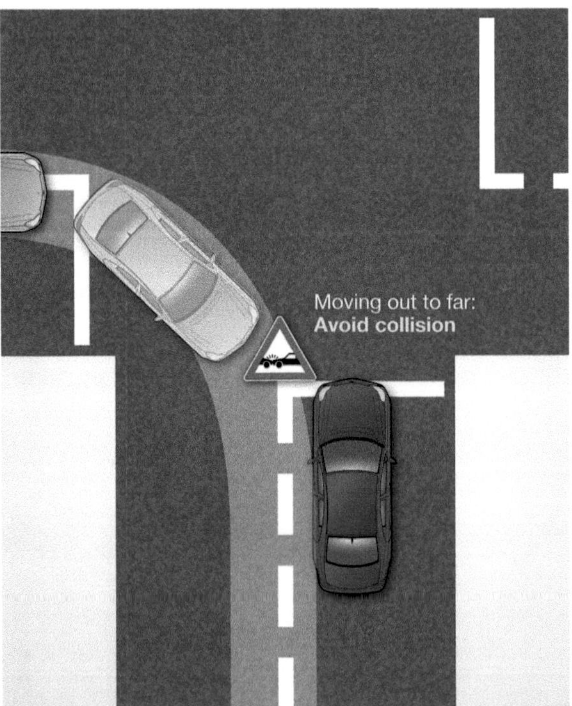

**Fig. 2.12** Curved
intersection

**Fig. 2.13** Roundabout
intersection

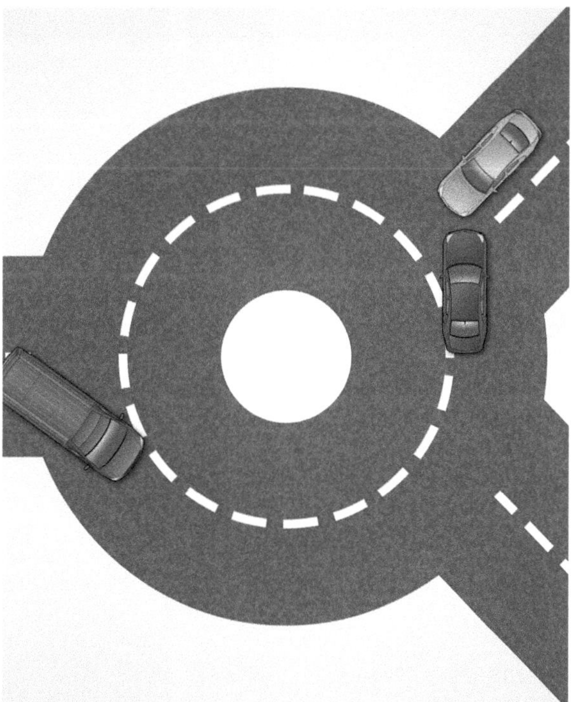

**Fig. 2.14** Merge lane

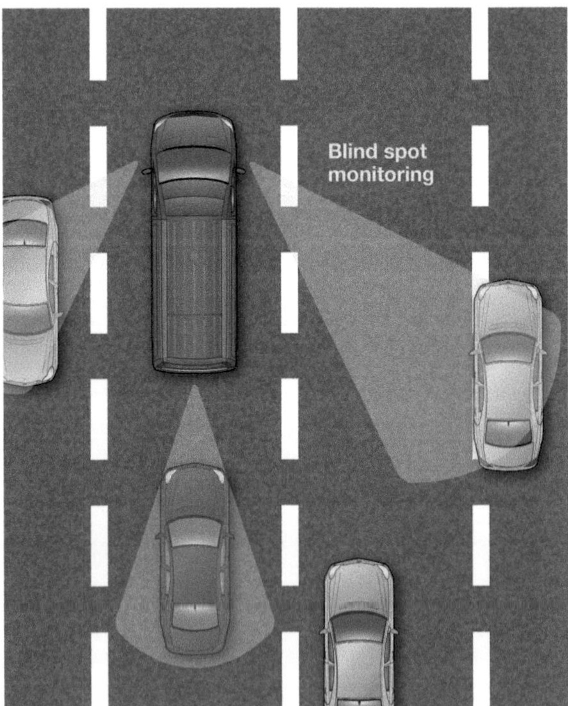


**Fig. 2.15** Blind spot areas
for a truck [14]

**Table 2.2**  Accident case study [6]

|            | At junctions (%) | Not at junctions (%) | Not defined (%) |
|------------|------------------|----------------------|-----------------|
| Car        | 17               | 77                   | 6               |
| Pedestrian | 24               | 70                   | 6               |
| Motorcycle | 28               | 68                   | 4               |
| Moped      | 34               | 62.5                 | 3.5             |
| Bicycle    | 44.5             | 52                   | 3.5             |
| Lorry      | 14.5             | 79                   | 6.5             |

According to the 2004 EU (Germany excluded) accident case study [6], only 21.7% of all accidents happened on a junction crossing . Almost 30% of all motorcycle and moped user fatalities occur at a junction crossing. In comparison, for car passengers only 17% of the fatalities happen at intersections. A complete Table 2.2 is presented below:

Note that most junction accidents (about 80%) occur inside of cities.

## 2.1.2  Classification of Safety Applications

Based on the usage of the applications we divide the safety domain in passive safety, active safety as well as proactive safety and warning applications (Fig. 2.16).

**Fig. 2.16**  Safety applications

Passive safety refers to features that help people stay alive and uninjured in a crash, such as seat belts and airbags. Size of the vehicle (bigger is safer) is also considered a passive safety feature.

Active safety functions help drivers to avoid accidents. This includes topics such as safe speed, driver monitoring, ABS brakes, brake assist (BA), electronic stability control (ESP), innovative assistance systems based on sensors technologies such as lateral support, night vision, and radar based safe following (automatic cruise control).

However, passive and active safety applications do not use vehicular communication and will therefore not be discussed further.

Proactive safety and warning based applications, on the other hand, make the object of this book and they will be detailed along with specific features (known as use cases).

Proactive safety features anticipate critical situations and take positive actions to deal with them, thus preventing possible accidents. This includes autonomous systems (such as collision avoidance) and non-autonomous systems (such as lane change warning). Proactive features are enhanced active safety features or brand-new features that would not be possible without vehicular communication. Still, active safety features should be encouraged, some, such as radar based should be used together with proactive safety features to check plausibility of information provided by the vehicular network. Proactive features are based on permanent beacons that contain telemetric data (position, velocity, etc) used to predict or prevent an event that may happen. Relevant examples of proactive safety applications are the ones that try to warn about a possible collision, such as cooperative forward collision warning, cooperative intersection warning or lane change assistance, further detailed in this section.

Warning features represent a temporary alert (a triggered event). We further classify these alerts into two types: quick and normal alerts.

The four groups of features (Autonomous & Non-Autonomous Systems and Quick & Normal Warning Alerts) that require vehicular communication are explained in detail below. Each group shares common requirements for more applications. The idea is to define general requirements for these groups, independent of individual applications that may arise in the future. Examples of individual applications will also be given.

*Autonomous Systems* do one or more tasks in an automated way. The full autonomous vehicle that will operate driverless with automated technology is probably about 10–15 years ahead, because of the multiple challenges. Besides the technological challenges, the social challenge is an important issue. The social aspect refers to issues such as to convince people to trust the vehicle. Legal issues also need to be resolved to allow the autonomous vehicles on public roads. On the technological level, previous efforts tried to create the autonomous vehicle based on information gathered from various sensors, such as cameras and radars. This was identified in projects from Italy (Argo) [15], European (Eureka Prometheus) [16] or USA (DARPA Grand Challenge) [17]. In the future, by using reliable

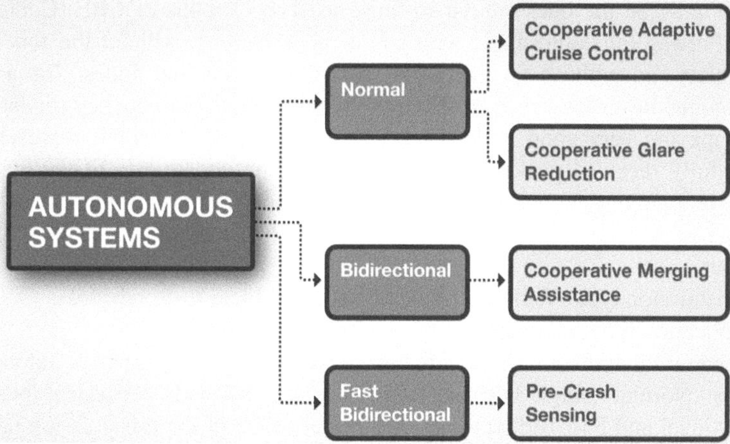


**Fig. 2.17**  Autonomous systems

vehicle communication with proper security, the autonomous vehicle will no longer be a dream, but reality. Of course, besides the proper communication, artificial intelligence inside the vehicle is required.

In this book we define autonomous systems as those based on message exchange to create automated tasks to enhance driving (such as cooperative adaptive cruise control) as depicted in Fig. 2.17. The function of the autonomous system is based on knowledge of its own status and on permanent beaconing information received from all surrounding vehicles in transmission range. In addition to wireless communications, object detection ("radar-based" sensors) has to be used as a backup.

The following requirements need to be met:

- a map data base (includes lanes geometry);
- permanent messages and specific sensors
- fusion from multiple sensors/vehicle communication;
- logic system that analyses incoming data and makes decisions.

The single-hop position based regime is achieved by the MAC layer as well as by a location discovery service and a location update service (explained in Sect. 3.2 and Chaps. 4 and 6). The discovery service only has to listen, as it is based on the information from neighbor nodes through the receipt of periodically sent out beacons, known as "permanent" beacons. Thus, an internal map with the single-hop neighbors of a vehicle can be built up. The location update has to keep two types of information for each node: radio range of the neighbors and single-hop neighbor nodes positions. In this case, no forward routing is required. The two services may be the ones similar to those used in Octopus or CBF (Contention-Based Forwarding) algorithms with slight modifications: without the forwarding mechanism and without specific reply from the surrounding nodes. The application requires lower latency than 100 ms since the message frequency (update rate) is 500 ms and very good PDR (Packet Delivery Ratio). The PDR represents the successfully received packets per sent packets. The update rate of the messages should be frequent enough not to influence the application instances, even if one message is skipped (lost). As identified in [18–20], a minimum PDR of 99% is necessary.

Transmission type: Vehicle-to-Vehicle and/or Vehicle-to-Roadside.

Application type: Proactive safety and Traffic Utilization.

Based on the transmission scheme used, we classify three types of autonomous systems: Normal (single-hop position based regime with permanent beacons), Bidirectional and Fast Bidirectional. We explain each of the transmission schemes in Chap. 6.

## 2.1.3 Normal Transmission Scheme

### 2.1.3.1 Cooperative Adaptive Cruise Control

Cooperative Adaptive Cruise Control (CACC) is a system that enhances the performance of adaptive cruise control (smoother braking and throttle, increase of traffic fluency and lower fuel consumption) by obtaining the dynamics of the lead vehicle and other vehicles ahead via beacon messages. Past [21] and current [22] research indicates that collision avoidance through automated cruise control is most suitable for significantly reducing the number of fatal accidents. The application improves safety as well as traffic efficiency on the road. Adaptive Cruise Control keeps a desired cruising speed, but also adjusts this speed, if required, to the vehicle in front, by braking. If required, it will stop the vehicle and then re-engage the throttle known as "stop and go". Then, it follows the front vehicle until that vehicle's speed is higher than the desired cruising speed or it just moves out of the lane. The desired cruising speed is set by the driver. The cooperative application should provide an optimum speed adaptation to follow the vehicle in front (if any) with the proper safety space between vehicles. It is a fully automated system and the driver is not required to interfere. The beacon messages are necessary because studies have shown that sensor-based systems sometimes have difficulties identifying which vehicles on a roadway are in the path of the host vehicle. The sensor systems (radar) can be particularly inaccurate during lane changes or where road segments change from straight to curved or vice versa. The vehicle communication enables a vehicle to follow its predecessor at closer distance under higher control. This is in part, because speed changes can be coordinated with each other [23, 24]. A simulation study shows that the number of shockwaves decreases with the rise of the CACC-equipped vehicles. Because close CACC platoons prevent other vehicles from cutting in, the system must coordinate with the Cooperative Merging Assistance in order to insert longer distance between vehicles or suggest a lane change in the merging areas. It lowers the negative effect of air pollution and energy consumption. Please note that although this application belongs mainly to proactive safety, it also improves traffic utilization & environment preservation.

### 2.1.3.2  Cooperative Glare Reduction

Cooperative Glare Reduction (CGR) is an automated system that, when darkness occurs, will switch from high-beams to low-beams and vice versa according to the distance to the surrounding vehicles in forward path area. Besides vehicle communication (establish the map of the surrounding vehicles), light sensors are required. The service may also be used for instance in intersections with no traffic lights to warn other drivers that come perpendicular. Because this application does not actually take control of the vehicle a PDR of 95% may be enough.

### 2.1.3.3  Cooperative Merging Assistance

Cooperative Merging Assistance (CMA) is a system that provides a safer, automatic way for a vehicle to join a flowing traffic (e.g. a highway entry). It allows vehicles to join ("on-ramp") the traffic without disrupting the flow of the traffic (Fig. 2.18). It eliminates the drivers misunderstandings by letting the vehicles decide the best way to join, based on the exchange of information (such as velocities and positions) between vehicles. It is obvious that a reliable automated system based on mathematical facts is much safer even compared with experienced drivers who may misjudge. This application improves safety, but also traffic efficiency on the road.

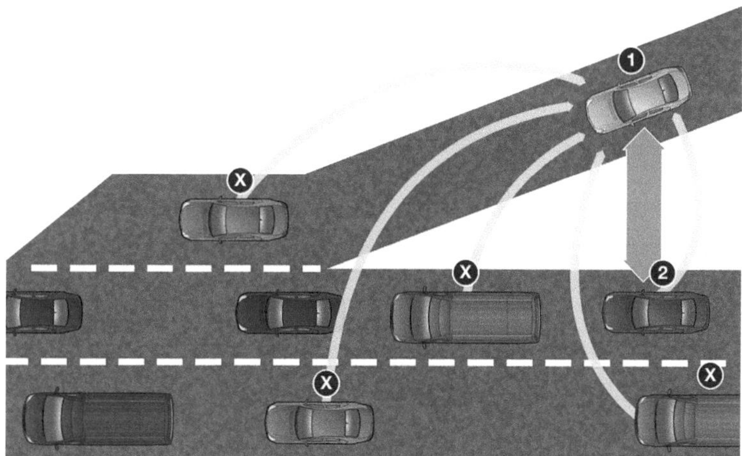

**Fig. 2.18**  Merging assistance

Misunderstandings between drivers on the highway and on the entry-lane can cause critical situations, especially when inexperienced drivers are involved.

The application consists in automatic traffic adjustment; it provides a ramp metering service where the merging vehicle is informed how fast it may proceed on the on-ramp in order to merge into an empty space in traffic [25]. By communicating, vehicles negotiate the merging process between each other. Particularly, suppose V1 wants to merge ahead of V2 based on prior query of all surrounding neighbors. Vx represents surrounding neighbors in one hop communication range. After the retrieve of position, speed and heading of surrounding vehicles (V2, Vx), V2 is identified by V1 as needed for a bidirectional connection. Therefore, a request of bidirectional regime with V2 is sent and an acknowledgement from V2 is expected. Then the actual connection maintenance (merge phase) is engaged and is ended by V1 when no longer required.

This particular application, which belongs mainly to proactive safety and secondly to traffic utilization, uses two transmission schemes: single-hop position based regime in order to gather telematics information about surrounding vehicles as well as bidirectional regime to negotiate the merging process with a vehicle in a particular position.

## *2.1.4 Bidirectional Transmission Scheme*

### 2.1.4.1 Pre-crash Sensing

Pre-Crash Sensing (PCS) is a system that optimizes safety of the drivers when an impact is imminent. Vehicles that are in unavoidable collision situations, based on beaconing information and radar-based sensors, engage a fast bidirectional unicast transmission with additional information like vehicle attributes for a better use of passive safety features, such as air bags, motorized seat belt pre-tensioners, and extendable bumpers (Fig. 2.19).

In particular, when the vehicle V1 (the initiator) detects an unavoidable collision with V2 (the responder), the vehicle logic of V1 requests a connection to V2. By

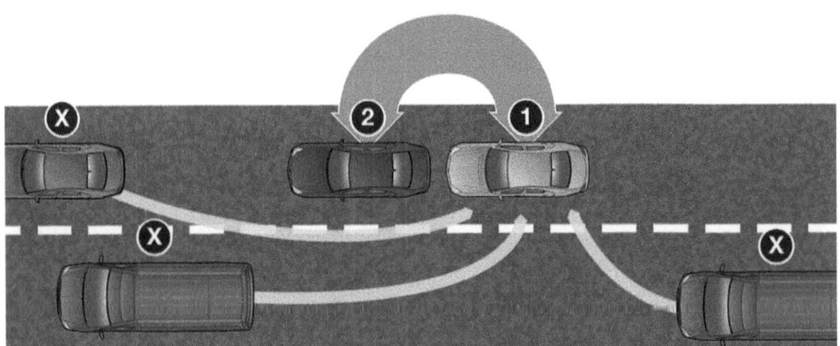

**Fig. 2.19** Pre-crash sensing

monitoring a list of the available services from the communication system, V1 is able to detect whether or not V2 supports the pre-crash sensing connection. If V2 supports pre-crash sensing, V1 sends a request to its communication system for the fast bidirectional connection. The communication system is assumed to make the connection according to lower-layer standards. Once the connection is established, the V1 requests for information (i.e., vehicle mass, bumper height) from the V2, that will provide the information. Vx represents surrounding neighbors in one hop communication range.

### 2.1.5 Non-autonomous Systems

*Non-autonomous Systems* are systems that prevent accidents with the help of an inside vehicle warn system (Fig. 2.20). They only announce the danger to the driver; they do not perform an automated task. These systems are based on one-way permanent beacon messages.

The inside vehicle warn system notifies the driver if a collision could occur. The decision to issue a warning is based on knowledge of its own status and on permanent beaconing information received from all surroundings vehicles in transmission range. When a vehicle makes a prediction of a collision, it warns the driver visually, auditory, and/or haptic. If the system expects that the driver still has enough time to avoid a crash, it will not attempt to control the vehicle. In addition to wireless communications, object detection ("radar-based" sensors) has to be used as a backup if communication fails or to identify vehicles that are not equipped with wireless communication.

The following requirements must be met:

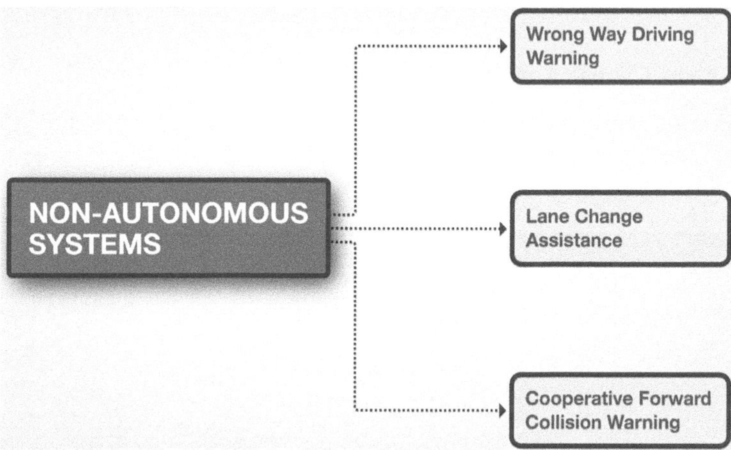


**Fig. 2.20**  Non-autonomous systems

- a map data base (includes lane geometry);
- permanent messages and specific sensors
- fusion from multiple sensors/vehicle communication;
- application logic that analyses incoming data and makes decisions to warn drivers;
- a Human Machine Interface (HMI), inside the vehicle, with visual, acoustic and haptic warnings.

Transmission type: Vehicle-to-Vehicle and/or Vehicle-to-Roadside.

Application type: Proactive safety.

Communication regime: Single-hop position based (with permanent beacons).

The single-hop position based regime with permanent beacons functions as described in Autonomous Systems paragraph with the mention that a minimum PDR of 95% is required.

Individual applications.

### 2.1.5.1  Cooperative Forward Collision Warning

Cooperative Forward Collision Warning (CFCW) is a system that warns the driver if a collision with a vehicle ahead may occur. Besides the awareness of the other vehicles in surrounding environment, communication can provide, for instance, sharing of the different "power" of breaking on each vehicle. This information becomes crucial if a larger vehicle such as a truck is involved. If a collision is imminent, the driver is warned to take appropriate action. It provides assistance to the driver primarily to avoid rear-end collisions with other vehicles. Rear-end collisions, all over the world, represent a significant percentage of all accidents. In particular, if V1 is the vehicle of interest. V2 is the vehicle ahead. Vx represent surrounding neighbors in one hop communication range. For V1 only messages from V2 are considered in the collision calculation (Fig. 2.21).

### 2.1.5.2  Lane Change Assistance

Lane Change Assistance (LCA) is a system that warns the driver in case he or she wants to change the vehicle's lane and other vehicles are approaching the respective

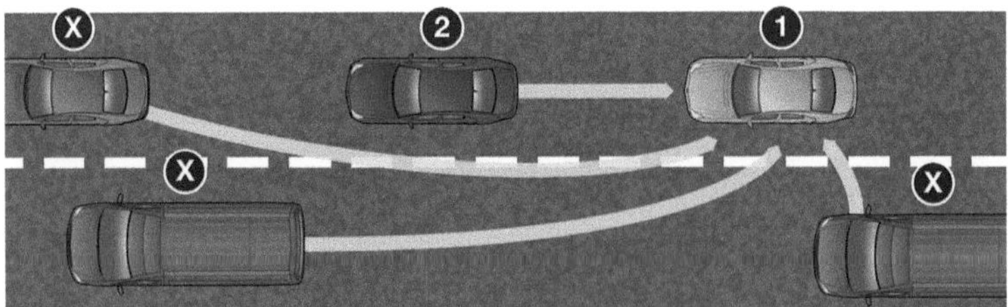

**Fig. 2.21**  Forward collision warning

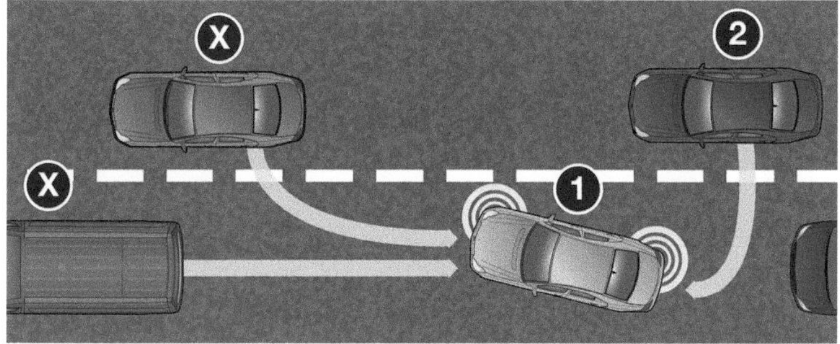

**Fig. 2.22**  Lane change assistance

**Fig. 2.23**  Wrong way driving warning

neighbor lane. This also incorporates a "blind spot" monitoring to the rear left/right of own vehicle and lane departure warning. The system prevents lateral related accidents, including blind spots, and assists the driver in bad visibility conditions (Fig. 2.22).

### 2.1.5.3  Wrong Way Driving Warning

Wrong Way Driving Warning (WWDW) is a system that identifies another vehicle driving wrong way and warns the driver of a possible collision (Fig. 2.23). Only the vehicles in forward path area are of concern. In Germany alone, radio stations report about 1,800 wrong-way drivers per year [26].

## 2.1.6  Quick Warning Alerts

Quick Warning Alerts are alerts that need to be sent to other vehicles as single hop, to a specific area if required (Fig. 2.24).

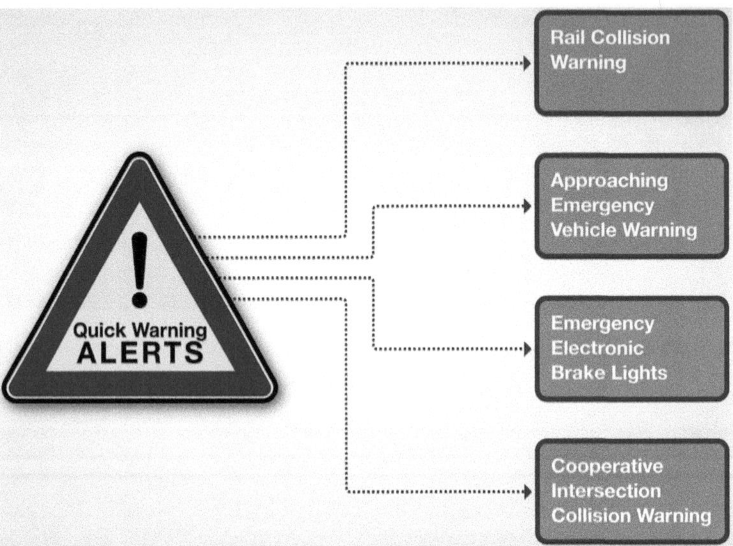


**Fig. 2.24** Quick warning alerts

The following specific requirements must be met:

- a map data base (includes lane geometry);
- alert messages;
- logic system that analyses incoming data and makes decisions to warn drivers;
- warning alerts are sent to vehicles accordingly;
- a Human Machine Interface (HMI), inside the vehicle, with visual, acoustic and haptic warnings.

Besides the alerts, permanent beacons are also required to create the table with the neighbors of a vehicle or roadside unit. This can be achieved using the single-hop position based regime with permanent beacons functions as described in Autonomous Systems paragraph.

The warning alerts are triggered by an event (a decision of a logic system placed on the roadside or on the vehicle). In the routing layer, the vehicles only listen the specific channel for warnings from roadside or other vehicles. The forward routing algorithm is not required. Location update and discovery is done through permanent beacons. Additionally, we try to define some particular values for the vital parameters in vehicle communication that are defined in Chap. 4. This application group requires a low latency because repetition of messages may be needed (see below), but not as low as the permanent beacon messages. In order to deliver the warnings fast enough, to provide good latency single-hop is used. Quick Warning Alerts require very good PDR, and here, a trick may be used to improve the PDR:

an alert is sent more than one time (repetition) to provide recovery from failure (lost packets). Two or even three repetition transmissions of the sender increase the PDR significantly. As identified in [18–20] a minimum PDR of 95% is necessary. The multiple warning (second and maybe third) must be in the so-called time to live (TTL) of the message. The TTL represents the time in which the warning is active, i.e. the time from when it is trigged until it expires (e.g. collision is inevitable or no longer occurs).

Transmission type: Vehicle-to-Vehicle and/or Vehicle-to-Roadside.
Application type: Warning.
Communication regime: Single-hop position based (with alerts).
Individual applications are described in the following subsections.

### 2.1.6.1  Cooperative Intersection Collision Warning

Cooperative Intersection Collision Warning (CICW) is a system that warns the driver if a collision in an intersection may occur. The system prevents vehicles from colliding based on the roadside unit warnings. Roadside units in intersections are used to coordinate between vehicles (assist the driver maneuver in intersection) and prevent violation of red light or stop sign. The decision of the roadside to issue a warning is based on permanent beaconing information received from all surrounding vehicles in transmission range. In this case permanent beacons are used to trigger the warning as well as to know where to send the warning. A detailed map of the intersection is required, providing elements such as: intersection orientation, number of lanes, stop bar location for all lanes. Roadside units monitoring the junction for approaching vehicles warn driver in case of unsafe situations, if e.g. a left-turning driver cannot see other approaching vehicles. In case of a traffic signal or stop sign violation the roadside unit has to detect and warn the driver in violation and, if necessary, other vehicles. The violation is issued based on the speed of the vehicles that need to stop. Based on the vehicle's location and speed (included in permanent beacons), the logic system determines whether a vehicle will cross into an intersection on red light or stop sign. Beside this, a warning should be issued if a vehicle is engaged dangerously in a left turn across path. These accidents represent about one third of the intersection related crashes and they occur because of failure to judge safe gaps (speeds of opposite path closing vehicles) correctly, as well as lack of driver's visibility.

### 2.1.6.2  Emergency Electronic Brake Lights

Emergency Electronic Brake Lights (EEBL) is a system that in case a vehicle has braked hard, it will send a warning to the vehicles behind it. The warning is not triggered by permanent information from surrounding vehicles, but permanent beacons are necessary only in order to know where to send the warning. False alarms need to be avoided.

### 2.1.6.3  Approaching Emergency Vehicle Warning

Approaching Emergency Vehicle Warning (AEVW). An emergency vehicle, when approaching other vehicles, sends a warning to vehicles in its path to move out of the way and free some emergency corridors. As identified in [8] this is known in specialized literature as blue corridor or blue wave. Here, the same message is repeated many times as long as vehicles are on the forward path.

### 2.1.6.4  Rail Collision Warning

Rail Collision Warning (RCW). A train when crossing a vehicle path, it sends warning to surrounding vehicles. This can be extended between trains. Here, the same message is repeated many times as long as the train is crossing a vehicle's road.

*Normal Warning Alerts* are alerts that need to be sent to other vehicles multi-hop in a time window, but not as fast as in single hop case, to a specific area if required (Fig. 2.25). These alerts announce an event that happened (e.g.: accident). Permanent beacons are also required to create the table with the neighbors of a vehicle or roadside unit.

The following specific requirements must be met:


- a map data base (includes lane geometry);
- alert messages
- logic system that analyses incoming data and makes decisions to warn drivers;
- warning alerts are sent to vehicles accordingly;

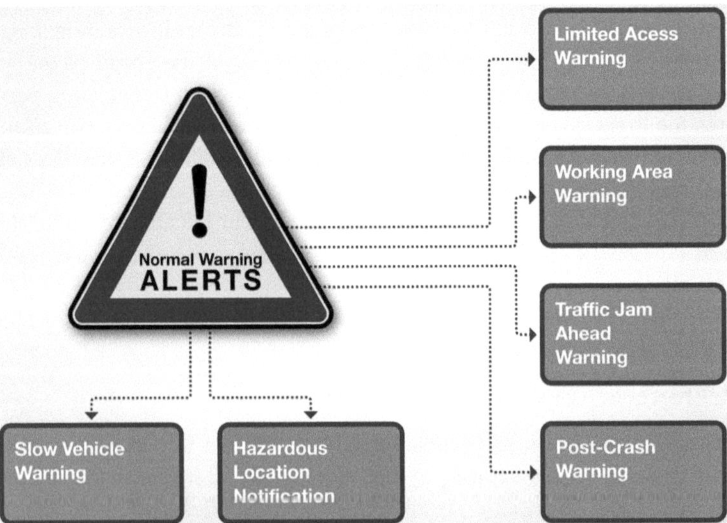


**Fig. 2.25**   Normal warning alerts

- a Human Machine Interface (HMI), inside the vehicle, with visual, acoustic and haptic warnings.

The main difference from single-hop alerts is that here, a forward routing algorithm is required. The challenge is to pick the best routing algorithm. As identified, there are many routing algorithms (see Chap. 4), but many of them do not provide scalability and reliability (many were used only in simulations) to meet the requirements of the vehicular communication. Because the TTL (time to live) of a message is much longer (up to hours or even days in case of working zone), the application does not require very low latency. The important thing is that the message reaches its destination (very good PDR), even at far distance; time is not that critical.

The application group requires very good PDR, and here as before, to improve the PDR: an alert is sent more than one time (repetition) to provide recovery from failure (lost packets). Two or even three repetition transmissions of the sender increase the PDR significantly.

Transmission type: Vehicle-to-Vehicle and/or Vehicle-to-Roadside.

Application type: Warning.

Communication regime: Multi-hop position based.

Individual applications.

### 2.1.6.5  Slow Vehicle Warning

Slow Vehicle Warning (SVW). Slow vehicles send warning alerts to other surrounding rear vehicles about their low speed in order to prevent collisions.

### 2.1.6.6  Post-crash Warning

Post-Crash Warning (PCW). Unmoving vehicles (because of an accident or mechanical failure) that disturb or endanger traffic send alert type warning messages to prevent collisions. Depending on the scene, the appropriate broadcasting scheme is selected, i.e.: on a highway geocast is used, as only rear vehicles need to be informed; on a regular road, multicast is needed, as it may endanger the vehicles on the other way as well.

### 2.1.6.7  Traffic Jam Ahead Warning

Traffic Jam Ahead Warning (TJAW). Vehicles in a traffic jam send warning alerts about the position of the jam end to approaching vehicles in order to avoid possible collisions. The electronic assistant informs the driver about the imminent situation, so he can slow down the car long before the danger comes into sight.

### 2.1.6.8  Hazardous Location Notification

Hazardous Location Notification (HLN). Send warning alerts about possible hazards detected by in-vehicle sensors (e.g. rollover, airbag deployment) or infrastructure

(RSUs). This information may be water or oil on a specific lane in a specific direction. Many hazard types may occur. For instance, an accident may induce fire in a vehicle [8]. Particularly in risky areas (e.g. in a tunnel), the concerned vehicle has to start the broadcasting of a local danger warning as soon as an excessive temperature is detected. Other hazard that may occur is slippery roadways. An estimation of road adhesion and slip requires information such as speed and acceleration of wheels. A vehicle may detect via ESP this road condition and make note of his location (latitude + longitude) as "dangerous". It stores and forwards information to other vehicles accordingly. It warns the driver and suggests a proper speed if needed. The hazard may also be transmitted / retransmitted by a roadside unit.

### 2.1.6.9 Working Area Warning

Working Area Warning (WAW). Send warning alerts about road works, blocked lanes announced by infrastructure (roadside units) or distributed by other vehicles.

### 2.1.6.10 Limited Access Warning

Limited Access Warning (LAW). Uses infrastructure to send warning alerts to the driver about restricted access such as height or weight (on a bridge for instance).

## 2.2 Resource Efficiency

Resource Efficiency domain captures issues regarding traffic congestion and fuel consumption.

Already in 1993 [21], estimations of the possible impact to traffic efficiency were conducted. Recently, a survey of telematics users in Japan depicted in [27] showed that road traffic information is most used and valued information for advanced telematics.

But why do we need traffic efficiency? Today, drivers from almost any major city face the same problem: everyday headache of traffic congestion. But the traffic congestion occurs not only in the major cities, but also on highways or roads between them. This has become a major problem of modern societies. For a few years already, in Europe, about 7,500 km or 10% of the road network are affected daily by traffic jams (European Commission, 2001) [6]. Congestion causes an increase of air pollution, of everyday stress degrading the quality of our lives. Crucial aspects of traffic efficiency solutions are themes related to cooperative vehicle-highway systems and automated vehicle-highway systems.

Simulation studies [28] show that traffic stability is improved by vehicle communication in terms of reduction in the number of shockwaves. Shockwaves occur in traffic flow propagate along a line of vehicles whenever traffic conditions change in front of the line, such as collisions, sudden decrease or increase in speed.

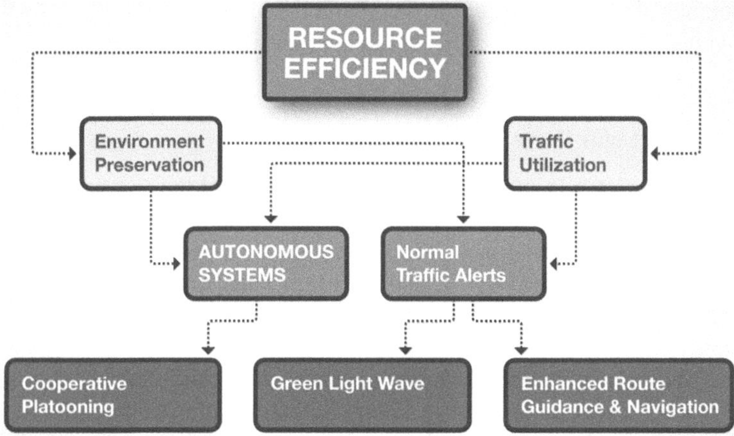


**Fig. 2.26** Resource efficiency applications

A classification of resource efficiency applications is depicted in Fig. 2.26.

Traffic utilization of roads are features that increase traffic fluency, such as traffic jam detection and avoidance, shortened travel times through cooperate and continuous driving. Generally, providing better traffic utilization will provide lower fuel consumption and therefore environment preservation.

Environment preservation refers to measures like fuel-efficient driving and reducing the general pollution caused by vehicle traffic.

### 2.2.1 Autonomous Systems

Cooperative Platooning (CP) refers to vehicle platoons (generally trucks) in which only the leading vehicle is driven by a human driver, the others are electronic controlled to close, but safe, follow the leader. This application improves traffic efficiency on the road and offers better reliability if exchange of information between vehicles is provided. It is an application about controlling the exact position, speed and acceleration of vehicles in a platoon, where only the leader vehicle has a human driver. This application belongs to traffic utilization & environment preservation.

### 2.2.2 Normal Traffic Alerts

Normal Traffic Alerts are sent like Normal Warning Alerts from safety domain; only the messages do not contain warnings, but information about the traffic. They have lower priority than normal warning alerts. Same requirements need to be met.

Transmission type: Vehicle-to-Roadside.

Application type: Traffic utilization.

Communication regime: Multi-hop position based.

Individual applications are described in the following subsections.

### 2.2.2.1  Green Light Wave

The infrastructure sends information about the position of the next intersection and the green signal time "window". Based on this information, the vehicle computes the necessary vehicle speed in order to catch the green light in the next intersection, so it does not have to stop. This results in a better traffic flow and lower fuel consumption, which saves time and money. This can be very efficient especially for commercial / logistic lorries. It consists of a smoother driving with fewer stops. Based on vehicular communication, the most advanced traffic control management system can help optimize the flow of traffic. Communication between cars and traffic lights generate flexible green waves on demand. For the system to work, the traffic light has to transmit its position, traffic signal phase and timing information for each direction. This application also belongs to environment preservation group.

### 2.2.2.2  Enhanced Route Guidance and Navigation

Enhanced Route Guidance and Navigation has the scope to reroute itinerary for navigation purpose. It refers to the ability of dynamic rerouting (of the path to the destination) based on up-to-date information about traffic flow from a traffic center. The vehicle uses this information to inform the driver about expected delays or better routes that might exist due to the traffic conditions. A service called Traffic Message Channel (TMC) over FM RDS already exists. A GPS navigation system connected to TMC is able to calculate a route in order to avoid congestions. Other services, like "TomTom Real-time traffic" that use an internet connection (GPRS) are also available. The main drawback of GPRS based systems is its costs: first you have to pay for the internet traffic (to mobile phone provider) and second a subscription may be required (e.g. to TomTom) in order to receive the traffic related information. Todays "real-time" updates from the traffic center occur about every 10–15 min, which is not enough. TMC also takes 30 s for a message from the traffic center to reach a car. The advantage of vehicular communication is that the information will be disseminated faster and the "real-time" database more often updated, based on other cars knowledge. But the most important advantage (which is not possible over FM RDS): different instructions may be sent to particular vehicles in a specific area (geocast). Otherwise, based on the information disseminated and on each vehicle's intelligence, all the vehicles would use the same new "best" route, creating a new jam. We give a concrete example: the highway is jammed. We need to send only a part of the vehicles on a different route, to avoid creating a new jam. If we may address vehicles in a particular area, the problem is solved. To conclude, in order to avoid congestion, [29–31] roadside units collect telemetric data from vehicles (via permanent beacons). Then, the data is sent to a central instance (vehicle-to-backoffice) and analyzed. Based on it (traffic fluency), optimum paths are computed and multi-hop disseminated, in a specific area, back to the vehicles.

## 2.3  Infotainment

Infotainment and Advanced Driver Assistance Services (ADAS), cover all additional features needed by drivers or passengers for a convenient travel as depicted in Fig. 2.27.

In-car services refer to features such as in-car entertainment (music, movies, games). These features do not use vehicular communication and therefore are not further detailed.

Ad-hoc services refer to temporary features, established for a specific duration, such as automatic toll payment.

Provider services refer to Internet access and other provider services such as service requirement notification. The vehicle drivers and passengers are thus able to browse the web while on the move, shop online, and even participate in videoconferences.

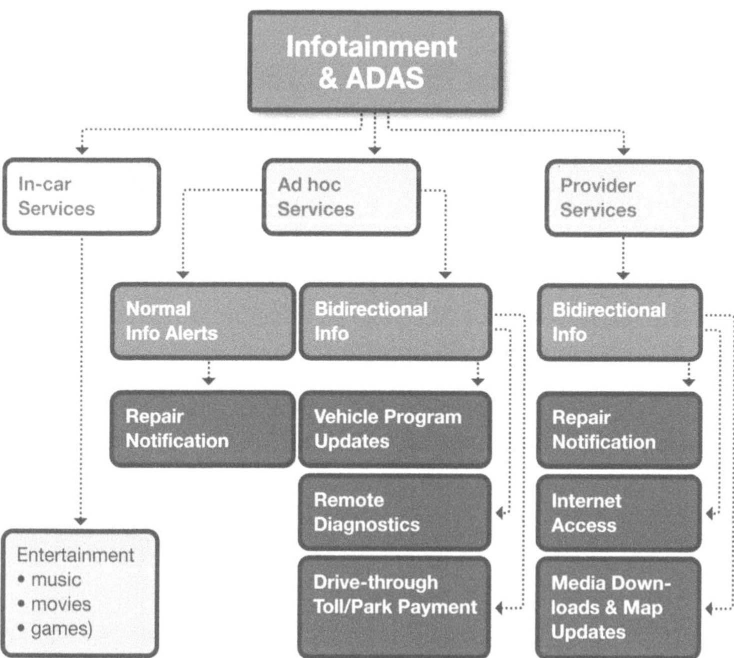


**Fig. 2.27**  Infotainment applications

### 2.3.1 Ad Hoc Services

Normal Info Alerts are sent like Normal Warning Alerts from safety domain; only the messages do not contain warnings, but information about the environment. They have lower priority than normal traffic alerts. Same requirements are needed.

Transmission type: Vehicle-to-Roadside.

Application type: Ad hoc services.

Communication regime: Multi-hop position based.

Bidirectional Info is sent like the one used by the applications of the Autonomous Systems with bidirectional transmission scheme (Sect. 2.1.2); only the messages are not safety related and therefore they have lower priority. No particular requirements are needed.

Transmission type: Vehicle-to-Roadside.

Communication regime: Bidirectional.

The individual applications for Ad hoc services are outlined in the following subsections.

#### 2.3.1.1 Point of Interest (PoI) Notification

Point of Interest (PoI) Notification uses infrastructure to send updated information about points of interest in the surroundings. The Point of Interest Notification allows tourist attractions, accommodations, shops, gas stations and other points of interest to announce themselves when a vehicle drives by. A roadside unit may broadcast additional information such as hours of operation and pricing. In a particular case of a restaurant, the actual menu of the day may be broadcasted. Although such POI already exists in different navigation software, they are "static" updated. In this scenario the benefit to consumers is up-to-date information from a business whose geographical location is near.

#### 2.3.1.2 Drive-Through Toll/Park Payment

*Drive-through Toll/Park Payment* refers to toll or parking space payment without the need to stop. This is already a fairly mature technology that allows for electronic payment of highway tolls [32]. An electronic monetary transaction occurs between a vehicle passing through a toll station and the toll agency. Electronic Toll Collection (ETC) systems require vehicle detection and classification and enforcement technologies. The ETC equipment substitutes for having a person (or coin machine) manually collecting tolls at toll booths. In addition, it allows these transactions to occur while vehicles travel at highway cruising speed.

#### 2.3.1.3 Remote Diagnostics

Remote Diagnostics allow a service station to assess the state of a vehicle, using vehicle communication. The service garage can query the vehicle for its diagnostic information to support the diagnosis of the problem reported by the customer. It can also retrieve vehicles' past history information from a database. The system may install software updates, if required. This application would reduce the time necessary for a repair, as well as lower the costs.

## *2.3.2  Provider Services*

Individual applications for Provider services (additional use of vehicle-to-backoffice):

### 2.3.2.1  Internet Access

Internet Access in vehicles allow an Internet connection in the car. A user can execute Internet-based applications from vehicle such as web browsing or email. The OBUs (on board units) may also provide [9] exchange of information between the vehicle and other units such as laptops, smart phones or PDAs via WiFi or Bluetooth. Using devices that already exist, is an advantage because the user is used to them and may access his own personal settings. From the vendor side, beside the one-time profit of selling the hardware, the subscriptions and pay per use for Internet access is another long-term profit. Over Internet or independent from it several other applications may be provided. Some examples are: Media Download to download audio, video data and applications; Map Downloads and Updates.

### 2.3.2.2  Repair Notification

Repair Notification sent in-vehicle diagnostics to a service center, advice and provides location of a nearby service to the driver if required.

### 2.3.2.3  Fleet Management

Fleet Management refers to the management of a fleet of vehicles by sending and receiving vehicle status data, transmitting mission data to the vehicle, returning mission status. This refers generally to commercial trucks but is not limited to it. The objective is to reduce overall transportation costs by improving efficiency and productivity. The key functions of the fleet management include vehicle tracking and diagnostics as well as driver and fuel management.

## 2.4  Summary of Application Requirements

In order for the above applications that require communication to be functional, we need to define the communication regimes required by these, which will be presented in the next section.

But, before that, to define a clear requirement of each application we present in Table 2.3 all the applications presented above and their requirements.

**Table 2.3** Applications requirements

| Applications | Comm. regime | Message type | Time critical | Transm. type | App. type | Priority | PDR (%) | Laten. (ms) | Sensor fusion | HMI | Map data |
|---|---|---|---|---|---|---|---|---|---|---|---|
| Cooperative glare reduction | SH | Beacons | RT | V2V | Proact | 0 | >95 | <100 | Yes | No | Yes |
| Cooperative adaptive cruise control | SH | Beacons | RT | V2V | Proact, traffic and environ. | 0 | >99 | <100 | Yes, for backup | No | Yes |
| Pre-crash sensing | SH, FB | Beacons | RT | V2V | Proact | 0 | >99 | <100 | Yes | No | No |
| Cooperative merging assistance | SH, B | Beacons, normal | RT | V2V | Proact and traffic | 0 | >99 | <100 | Yes, for backup | No | Yes |
| Cooperative platooning | SH, B | Beacons, normal | RT | V2V | Traffic | 0 | >99 | <100 | Yes | No | Yes |
| Cooperative forward collision warning | SH | Beacons | RT | V2V | Proact | 0 | >95 | <100 | Yes, for backup | Yes | Yes |
| Lane change assistance | SH | Beacons | RT | V2V | Proact | 0 | >95 | <100 | Yes, for backup | Yes | Yes |
| Wrong way driving | SH | Beacons | RT | V2V | Proact | 0 | >95 | <100 | Yes, for backup | Yes | Yes |
| Cooperative intersection collision warning | SH | Beacons, alerts | RT | V2V, V2R | Warning | 1 | >95 | <200 | No | Yes | Yes |
| Emergency electronic brake lights | SH | Beacons, alerts | RT | V2V | Warning | 1 | >95 | <200 | No | Yes | Yes |
| Approaching emergency vehicle warning | SH | Alerts | TTL | V2V | Warning | 1 | >95 | <200 | No | Yes | Yes |
| Rail collision warning | SH | Alerts | TTL | V2V | Warning | 1 | >95 | <200 | No | Yes | No |
| Slow vehicle warning | MH | Alerts | TTL | V2V | Warning | 2 | >95 | <400 | No | Yes | Yes |

**Table 2.3** (continued)

| Applications | Comm. regime | Message type | Time critical | Transm. type | App. type | Priority | PDR (%) | Laten. (ms) | Sensor fusion | HMI | Map data |
|---|---|---|---|---|---|---|---|---|---|---|---|
| Post-crash warning | MH | Alerts | TTL | V2V | Warning | 2 | >95 | <400 | No | Yes | Yes |
| Traffic jam ahead warning | MH | Alerts | TTL | V2V | Warning | 2 | >95 | <400 | No | Yes | Yes |
| Hazardous location notification | MH | Alerts | TTL | V2V, V2R | Warning | 2 | >95 | <400 | No | Yes | Yes |
| Working area warning | MH | Alerts | TTL | V2V, V2R | Warning | 2 | >95 | <400 | No | Yes | Yes |
| Limited access warning | MH | Alerts | TTL | V2R | Warning | 2 | >95 | <400 | No | Yes | Yes |
| Green light wave | MH | Alerts | TTL | V2R | Traffic and environ. | 3 | >95 | <400 | No | No | Yes |
| Enhanced route guidance & navigation | MH | Beacons, alerts | TTL | V2R, V2B | Traffic | 3 | >95 | <400 | No | Yes | Yes |
| Point of interest notification | MH | Alerts | TTL | V2R | Ad-hoc | 4 | >95 | <400 | No | Yes | No |
| Drive-through toll/park payment | B | Normal | TTL | V2R | Ad-hoc | 4 | >95 | <400 | No | Yes | No |
| Remote diagnostics | B | Normal | TTL | V2R | Ad-hoc | 4 | >95 | <400 | No | Yes | No |
| Internet access | B | Normal | TTL | V2R, V2B | Provider | 4 | >95 | <400 | No | Yes | No |
| Repair notification | B | Normal | TTL | V2R, V2B | Provider | 4 | >95 | <400 | No | Yes | No |
| Fleet management | B | Normal | TTL | V2R, V2B | Provider | 4 | >95 | <400 | No | Yes | No |

Abbreviations: B – Bidirectional; FB – Fast Bidirectional; SH – Single-hop position based; MH – Multi-hop position based; RT – Real-time; TTL – Time To Live (Time constrain); Priority – 0 Highest, 4 Lowest; HMI – Human Machine Interface (audio, visual, haptic); Normal – messages that are exchanged in bidirectional regime.

# References


1. http://www.ami-c.org/
2. World report on road traffic injury prevention, World Health Organization, 2004
3. http://www.bicyclesource.com/body/safety/collision-types-occurence.shtml
4. K. Langwieder, Car crash collision types and passenger injuries in dependency upon car construction (field studies of the German Automobile Insurance Companies)
5. "Road traffic accident", March 2007, http://en.wikipedia.org/wiki/Road_traffic_accident
6. http://ec.europa.eu/transport/roadsafety/road_safety_observatory/care_en.htm
7. L. Andreone and M. Provera, Inter-vehicle communication and cooperative systems: local dynamic safety information distributed among the infrastructure and the vehicles as "virtual sensors" to enhance road safety, http://www.car-to-car.org
8. G. Segarra, Activities and applications of the car 2 car communication: the renault vision, http://www.car-to-car.org
9. K. Matheus, R. Morich, I. Paulus, C. Menig, A. Lübke, B. Rech, and W. Specks, car-to-car communication – market introduction and success factors, Germany
10. http://www.stvincent.ac.uk/Resources/Physics/Speed/speed/index.html
11. http://www.ae-plus.com/Key%20topics/kt-brakes-news6.htm, Brakes & Steering, August 2005
12. http://www.carsense.org/
13. A. Hoess, Multifunctional automotive radar network (RadarNet) final report, RadarNet consortium, 2004
14. http://homesteadschools.com/traffic/course/
15. http://www.argo.ce.unipr.it
16. "EUREKA Prometheus Project", March 2008, http://en.wikipedia.org/wiki/EUREKA_Prometheus_Project
17. DARPA Urban challenge 2007, http://www.darpa.mil/grandchallenge/, accessed March 2008
18. M. Torrent-Moreno, F. Schmidt-Eisenlohr, H. Füßler, and H. Hartenstein, Packet forwarding in VANETs, the complete set of results, Karlsruhe, 2006
19. S. Yousefi, A. Benslimane, and M. Fathy, Performance of beacon safety message dissemination in vehicular ad hoc network
20. X. Ma and X. Chen, Performance analysis and enhancement of safety applications in DSRC vehicular ad hoc networks, 2007
21. P. Varaiya, Smart cars on smart roads: problems of control. IEEE Transactions on Automatic Control, Vol. 38, No. 2, pp. 195–207, February 1993
22. P. Varaiya, V2V (vehicle to vehicle) and V2I (vehicle to infrastructure) communication: does the technology match the problem? In Proceedings of ESCAR 2006 Conference Programm, Berlin, November 2006
23. A. Vahidi and A. Eskandarian, Research advances in intelligent collision avoidance and adaptive cruise control. IEEE Transactions on Intelligent Transportation Systems, Vol. 4, No. 3, pp. 143–153, September 2003
24. B. van Arem, C.J.G. van Driel, and R. Visser, The impact of cooperative adaptive cruise control on traffic-flow characteristics, 2006
25. http://www.cartalk2000.net/
26. http://www.ertico.com
27. M. Matsumoto, Japanese new challenge for mobile office services in the DSRC. The fully networked car workshop, Geneva, Italy, March 2007
28. K.M. Malone and B. Van Arem, Traffic effects of inter-vehicle communication applications in CarTALK 2000. Paper presented at the 11th World Congress on ITS, Nagoya, Japan, 2004
29. http://www.tmcforum.com
30. http://www.tomtom.com/
31. Car2Car Communication Consortium – Application Working Group, 2006, www.car-to-car.org
32. G.S. Bickel, Inter/intra-vehicle wireless communication, 2006

# Chapter 3
# Communication Regimes

Orthogonal to the application domains are the vehicle communication regimes. Classifying these regimes is multi-facetted. We define communication regimes according to the technology required (transmission scheme: bidirectional and position based), but we also define the regimes according to the usage (Fig. 3.1) (transmission type: in-vehicle, vehicle-to-vehicle, vehicle-to-infrastructure, vehicle-to-backoffice).

In-vehicle communication refers to the communication between different electronic units (like sensors and actuators) inside the vehicle. The inside vehicle communication is wired and several network types exist to best suit the different in-vehicle automotive applications. So, there is a low-cost, short-distance, low-speed network for non-critical applications such as control of seats or sunroofs called LIN (Local Interconnect Network). The most common in-vehicle network that provides a fast and reliable link between sensors and actuators is CAN (Controller Area Network) [1–3]. This bus supports automotive critical or non-critical applications. Some examples are active safety applications (ABS, ESP, parking aid).

Buses for particular applications are FlexRay and MOST. FlexRay [4–6] is designed especially for applications that require high-speed control such as: powertrain (engine control), x-by-wire (feedback von brake, throttle) but also active safety systems (ESP). MOST [1, 7, 8] is a multimedia fiber-optic network that provides high speed data transfer and a low cost interface for simple multimedia applications (microphones and speakers) as well as for more complex applications.

*Vehicle-to-Backoffice* refers to communication with existing infrastructure via standards, such as GSM or UMTS. As identified in specialized literature this is not suited for proactive safety applications. It may be used in infotainment (e.g. Internet Access) and particular applications of traffic utilization (Enhanced Route Guidance and Navigation).

*Roadside-to-Backoffice* also refers to the exchange of information between roadside units and the so-called backoffice that may use existing wireless infrastructure such as GSM, UMTS but also land line.

Satellite-to-Vehicle is actually a localization service of the vehicles that might use satellite-based systems such as global positioning system (GPS) or in future Galileo. Although, GPS alone does not provide enough accuracy and not enough frequency update. Therefore further research is required.

R. Popescu-Zeletin et al., *Vehicular-2-X Communication*,
DOI 10.1007/978-3-540-77143-2_3, © Springer-Verlag Berlin Heidelberg 2010

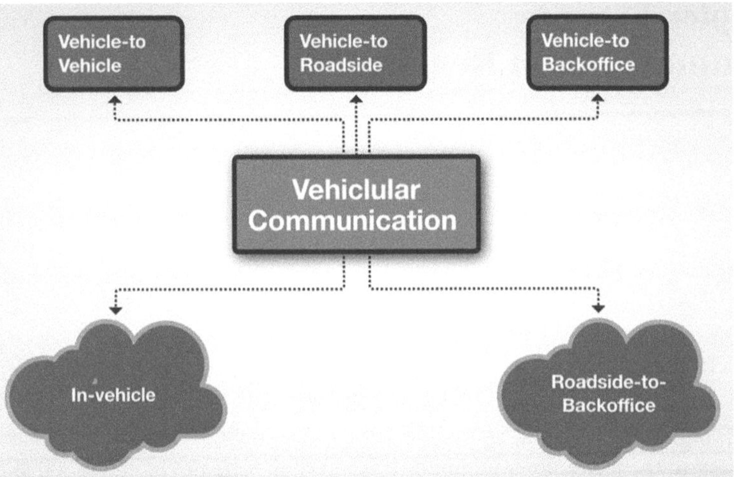


**Fig. 3.1**  Vehicle transmission types

The transmission types presented above refer to classic technologies already well-documented in the literature. Therefore, they are not further detailed in this book.

The book will study the two remaining transmission types in detail: Vehicle-to-Vehicle (V2V) and Vehicle-to-Roadside (V2R) communication. Both of these types have to use either a bidirectional or a position based regime in order to provide the features of the three application domains previously presented.

Some of the features such as cooperative platooning, toll payment or remote diagnostics need the bidirectional regime, but others such as cooperative forward collision warning, cooperative adaptive cruise control, or hazardous location notification need a special position based regime. Particular features, such as cooperative merging assistance, require both regime types.

The position-based regime is a must-have requirement for many of the features because they need addressing to vehicles in a particular geographical area.

## 3.1 Bidirectional Communication Regime

The bidirectional regime, also known as unicast, enables a connection between two vehicles or vehicle and roadside for bidirectional exchange of information (Fig. 3.2), where the roadside station can be interpreted as a fixed position vehicle. This also means that for each information sent, a respective acknowledge from the receiver is expected such that no loss of information occurs. This way, more reliability is provided.

The implementation of the bidirectional regime contains four phases. In the *Discovery* phase, one of the vehicles searches for surrounding nodes (another vehicle or roadside unit). In the *Connection* phase, a vehicle initiates a connection to

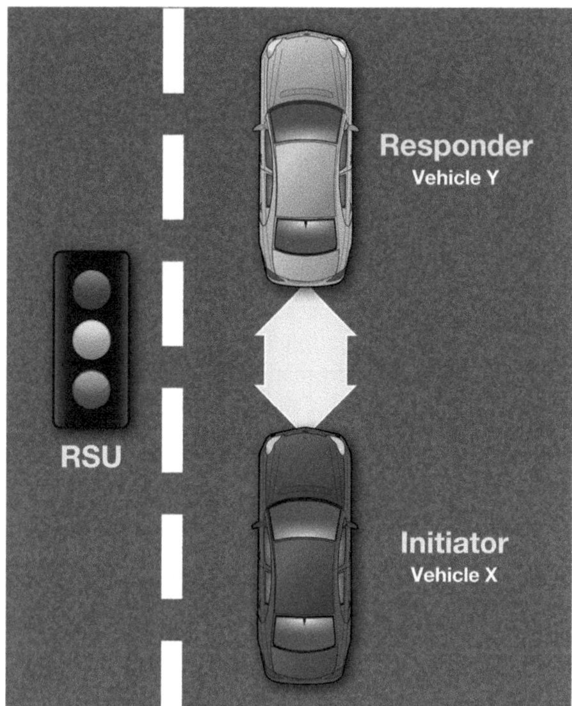



**Fig. 3.2** Example of bidirectional communication

another vehicle or roadside. The other vehicle/roadside unit is able to allow or deny the connection according to a set of rules. In the *Data* phase, the two parties keep the connection open while exchanging information. In the *Ending* phase, one of the two parties decides to end the connection and the communicating partners stop exchanging information.

There are some requirements for the above phases (such as opening of a different communication channel) that should be transparent to the applications and should be resolved by the lower communication layer.

The *Initiator* needs to:

- execute the discovery phase;
- identify and select a suitable vehicle/roadside with the specific service;
- send a request for a connection to the Responder
- execute a unicast, two-way communication between the two parties by exchanging messages at appropriate times.

The *Responder* needs to:

- reply to all connection requests (accept or refuse);
- authenticate and check plausibility of the messages from vehicle;
- execute an unicast, two-way communication between the two vehicles by exchanging messages at appropriate times.

The Initiator as well as the Responder may close the connection at any time.

There are some particularities when the connection occurs with a roadside as it may need to route messages to backoffice (i.e.: Internet Access). But this transmission can in fact be done over regular land lines, making no interest to our book.

The bidirectional regime has the advantage that two particular vehicles (or a vehicle and a roadside unit) can engage an exchange of information in both ways on demand. This exchange of information is always acknowledged by the other side, so that no loss of information occurs. Thus, it can be said that the regime offers the benefits of interactivity between the two parties.

But due to the bidirectional exchange of information and waiting for an acknowledgement after the information is sent, delays occur. A longer delay and more network load is needed if the information has to be transmitted to more than one vehicle. Therefore, this type of regime is not suited for some of the applications, but is a must for others.

## 3.2 Position Based Communication Regime

The position based regime is a particular mechanism in which the information is spread simultaneously only to a group of vehicles in a specified geographical area, also known in specialized literature as geocast (Fig. 3.3).

**Fig. 3.3** Geocast

The information is disseminated one way only in the network by vehicles or roadside units, where roadside is seen as a fixed position vehicle.

The implementation of the position-based regime contains two phases. The *Discovery* is the phase where one of the vehicles or roadside units decides to send information to other vehicles within a specific geographical area. The *Location Update* has to keep real-time position of surrounding neighbors. The *Flooding* is the phase where the participant delivers the information tagged (with the desired geographical area). The vehicle that receives the information checks the tag and keeps it or discards it accordingly (Fig. 3.4).

The instances of the communication regime are:

The *Sender* needs to:

- obtain the information (may be local vehicle telemetric data such as position or velocity or other stored information for a certain time)
- package the information data into a message;
- use a geocast mechanism to send the message to surrounding vehicles if any.

The *Receiver* needs to:

- obtain messages from sender;

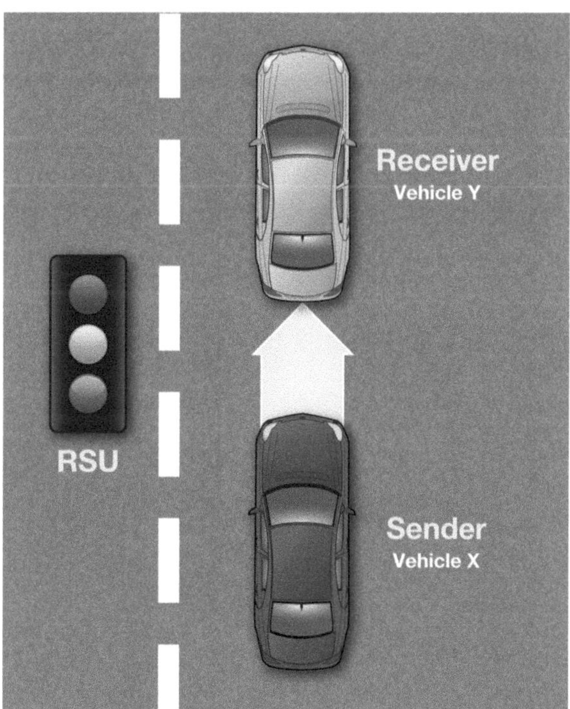


**Fig. 3.4** Example of position-based communications

- decode the messages into local vehicle data;
- check plausibility of the messages from other vehicle by comparing with local sensor data.

The most important advantage of this regime is that it transmits information to vehicles in a particular area, which is a must for some of the applications. The regime, having the ability to deliver information very fast to a large number of vehicles, also lowers network load and saves time for dissemination of information. The disadvantage is that it transmits information only one way, which means no interactivity with the other party and no confirmation that the information was successfully received.

## 3.3 Multi-Hop Position Based Communication Regime

But what if the information needs to travel from one vehicle to another and so on in a chain? Then the information has to take multiple hops to get to the destination (Fig. 3.5). In order to do this, a so-called routing algorithm is required in order to find the right next hop.

In position-based regime, routing requires a localization mechanism (physical position of each participant). The routing has to achieve two things. One, to determine the position of the destination (actually map a vehicle to its geographical position) and second to select one of its neighbors as the next hop the information should be forwarded to.

So, concluding we can say that according to the communication needs we have three communication regimes:

- Bidirectional (classic or fast, see Chap. 6 for more details),
- Single-hop position based (with one-way permanent beacon or alerts),

  Multi-hop position based (with alerts).

**Fig. 3.5** Multi-hop

# References


1. http://www.epanorama.net
2. D. Marsh, CANbus networks break into mainstream use, Europe: EDN, 2002, http://www.edn.com
3. http://www.can-cia.org
4. "FlexRay", December 2006, http://en.wikipedia.org/wiki/FlexRay
5. http://www.flexray.de
6. FlexRay, Communications system protocol specification version 2.1 revision A, FlexRay consortium, December 2005
7. http://www.mostcooperation.com/
8. MOST Cooperation, MOST specification revision 2.4, 2005

# Chapter 4
# Information in the Vehicular Network

We classify the required information into three categories: a general one common for any application and other two with particular information for permanent beacons and for alerts. Please note that permanent beacons are actually periodic but with high frequency rate and we call them permanent to distinguish them from alerts (see Chap. 6 for details).

Common information (encapsulated as a packet header) that needs to be transmitted in the vehicular network is composed of:

- msgID with timestamp;
- nodeID, that represents a unique ID, might be an IP;
- ownNodeType that represents vehicle or roadside.

Information that needs to be transmitted via permanent beacons in the vehicular network is composed of:

- telemetric data: position (current lane included), speed, acceleration heading and yaw-rate of all surrounding vehicles, including itself (discovery & location update);
- position confidence level that may be influenced by localization signal strength;
- internal vehicle parameters, such as: turn signal status (left, right or no signal), ABS, ESP status (denoting slippery road), brake response time (based on vehicle's mass, tire condition).

Besides the permanent messages, alerts are to be transmitted, with additional information such as:

- area with slippery roads;
- wind and speed direction;
- any other traffic related announcements such as hazard or working area;
- priority level of the information;
- time to live (TTL) of the information representing the time frame in which the alert is active;
- reliability of the information set by sender.

R. Popescu-Zeletin et al., *Vehicular-2-X Communication*,
DOI 10.1007/978-3-540-77143-2_4, © Springer-Verlag Berlin Heidelberg 2010

Besides the communicated information some applications require a digital map in order to provide road geometry with information such as number of lanes & elevation.

To represent the above information we require the use of vectors. The velocity vector, for instance, requires additional two bytes to store the information about nodes speed and direction. The first byte encodes the direction in the range of 0–127 (7 bits), where the Most Significant Bit (MSB) indicates whether velocity vector information is available or not. The second byte stores the speed in km/h. This however, would limit the representation of the maximum speed: 0–255.

For the permanent beacons, should be noted that the sampling frequency of the localization service should be shorter than the frequency of the messages being sent, otherwise there is the chance of data mismatch. The frequency of the messages being sent should be 2÷20 Hz (i.e.: 50÷500 ms) [1–3], while the GPS has a sampling frequency of 1 Hz, thus making the normal GPS system not suited.

For the alert type messages such a high frequency is not required, 2–30 s, depending on the type of alert (a vehicle behind a curve should send messages with a higher update rate than the traffic information) should be enough. The alert type messages are event based; they are valid for a specific period, a so-called time to live (TTL).


## 4.1 Accuracy of Information

The position, velocity, acceleration, heading, yaw-rate, pitch angle, pitch-rate, roll angle, roll-rate parameters may be "sensed" by an integrated inertial and GPS navigation system (e.g.: RT3002 reaching a position accuracy of 2 cm using RTK). To achieve an accurate relative position of vehicles (including driving lane) [4–7] only by the GPS device a Differential or RTK GPS/GLONASS correction service through an infrastructure network is required (or in future Galileo). At present, the GPRS/UMTS networks may be used, but require subscription/costs. They offer the possibility of bi-directional communication! (corrections will consist only of a subset of data, which helps to decrease the amount of information needed to be transferred through the GPRS network). Today, Real-time DGPS reaches a 20 cm accuracy and Real-time Kinematic (RTK) reaches as low as 1 cm accuracy! DGPS normally has an accuracy of 1–5 m (by using for example the closest fixed base station from Berlin: MAUKEN on Elbe), but with special type of base station (e.g.: Trimble Pro XR) can achieve a 30 cm accuracy for a minimum of 5 min and even 1 cm for a minimum of 45 min. With RTK, you need a base station placed on a known, surveyed point, and one or more mobile receivers within a 50 km range of your base station.

One important requirement is real-time update of information of the involved neighbors! Otherwise, the neighbor to whom the information is outdated may take an action that can cause bad consequences (e.g.: accident). A key factor is the vehicle's logic, the so-called in-vehicle "intelligence" [8, 9]. Based on the information received from the sender (e.g.: brake), it will act accordingly (e.g.: also brake).

## 4.2 Time Critical Information

We argue that the information for all applications (except maybe some from the infotainment domain) in vehicular communication is time critical. Time determines wether the information is meaningful or not. This is why we provide two concrete examples where time is critical. The most obvious examples are when a vehicle has to brake or to overtake (in areas with one lane per way). We show the time to brake according to surface conditions and external factors, vehicle technologies and vehicle weight. We point out that, if the information is not sent at the right time, there is no point in sending it anymore. In other words, if the message is not sent in real-time or time constrained, we conclude that the information sent is valueless.

Then, in Sect. 4.5 we define time zones that are critical in an advanced collision avoidance system based on vehicular communication. We present a table with time constrains for each of the zones according to the obstacle type. We conclude that communication-based collision avoidance system should provide better reaction time (faster response time) in terms of brake and evasive maneuvers. We also argue that, for advanced collision avoidance, we need both, vehicle communication and sensors. And that these two are interconnected and complementary to each other.

## 4.3 Time and Distance for Braking

The stopping distance of a vehicle is mainly influenced by the friction coefficient with the road. The friction ($\mu$) experienced values are [10, 11] (Fig. 4.1):

- Dry asphalt: 0.8–1.2
- Wet asphalt: 0.5–0.8
- Snow: 0.1–0.3
- Ice: 0.03–0.15

In order to stop a car, the kinetic energy must be reduced to zero or the kinetic energy must equal the energy given by the friction force [12, 13]:

$$F_f \cdot d = \mu \cdot m \cdot g \cdot d = \frac{1}{2}m \cdot v_0^2$$

(1) Friction force
The stopping distance is:

$$d = \frac{v_0^2}{2 \cdot \mu \cdot g}$$

(2) Stopping distance

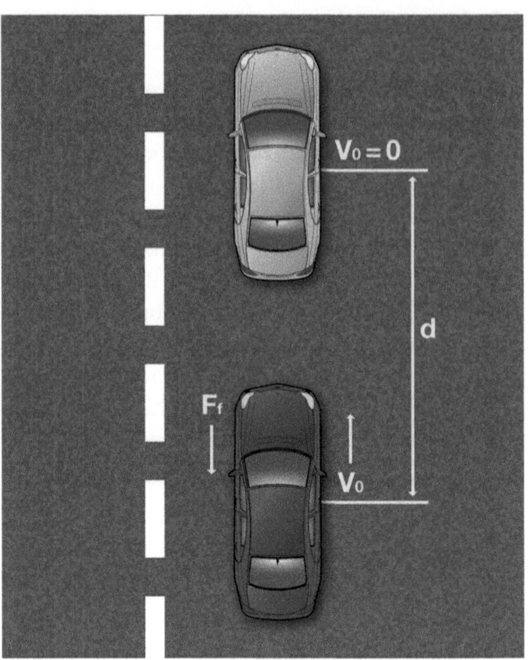


The above formula assumes a constant rate of deceleration of $\mu \cdot g$ (if $\mu = 1$ the deceleration is 9.8 m/s$^2$). This deceleration may be achieved under closed course conditions by professional drivers; a reasonably skilled driver could only get a deceleration of 6 m/s$^2$ without loss of control.

$$v^2 = v_0^2 + 2a(d - d_0)$$

(3) Galilei equation (/)

$$v = v_0 + a \cdot t$$

(4) Velocity equation (/)
Based on (4), (3), (2), the stopping time is:

$$t = \frac{v_0}{\mu \cdot g}$$

(5) Stopping time
The distance at constant velocity (no break) is:

$$d_2 = v_0 \cdot t$$

(6) Distance path with no break

Theoretical values of stopping distance and time with different friction coefficients are presented in the Figs. 4.2 and 4.3. These values are independent of vehicle mass, contact surface (tire width) and brake response time (human reaction delay).

The stopping time on a wet surface increases (see Figs. 4.4 and 4.5). But because of aquaplaning, the speed should be limited with these conditions (by external sensors) to a maximum of 160 km/h or even less.

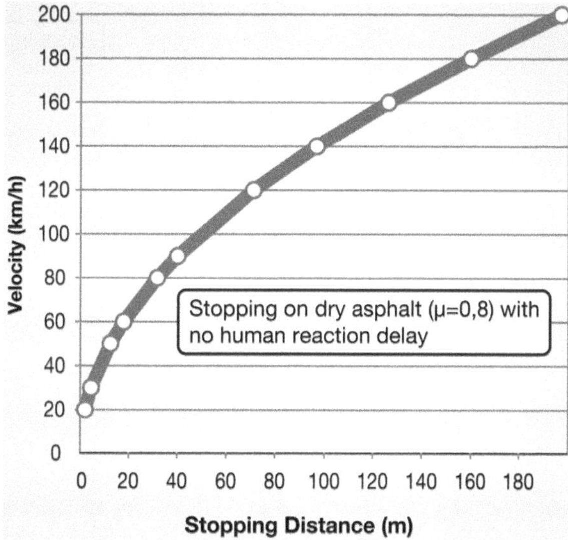


**Fig. 4.2** Stopping distance on dry asphalt ($\mu = 0.8$)

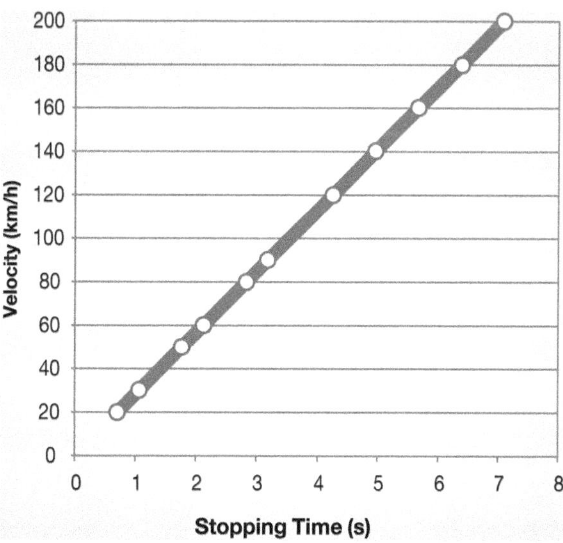


**Fig. 4.3** Stopping time on dry asphalt ($\mu = 0.8$)

**Fig. 4.4** Stopping distance on wet asphalt ($\mu = 0.5$)

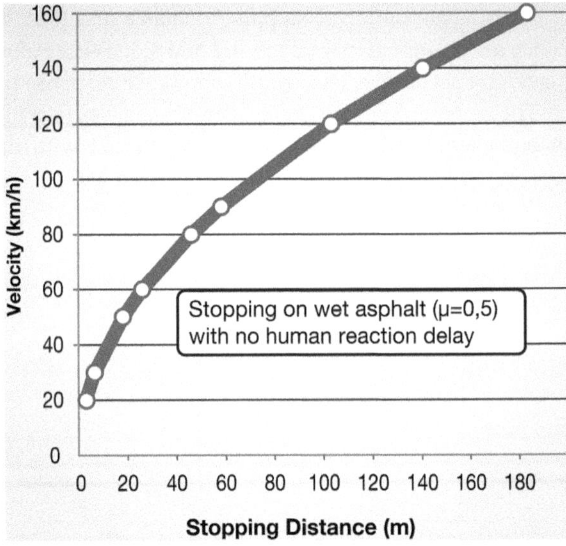


**Fig. 4.5** Stopping time on wet asphalt ($\mu = 0.5$)

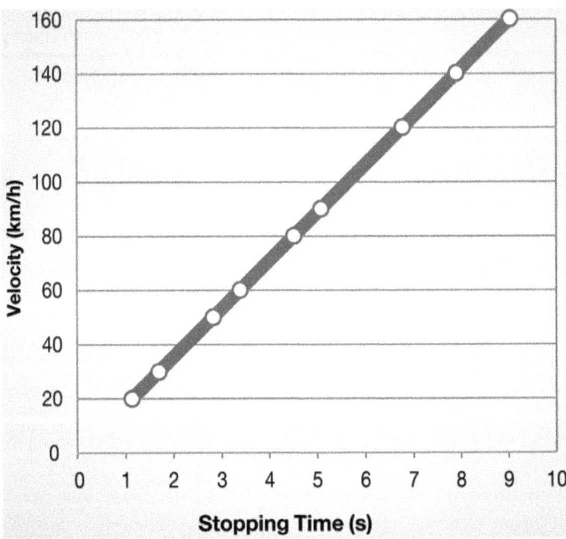


Aquaplaning [14, 10] starts to be non-negligible from speeds of v > 90 km/h (with new tires). The phenomenon represents hydrodynamic water buildup between the tire and the road surface. In order to have sufficient time for the water to be squeezed out of the contact regions between the tire and the road surface, the following relation should be fulfilled:

$$v < \left( \frac{\sigma}{\rho} \right)^{\frac{1}{2}}$$

(7) Aquaplaning speed
where:


$\sigma$ – Perpendicular stress in the tire-road contact area
$\rho$ – Water mass density


The stopping time on a snow surface increases dramatically. However, this effect is even worse on ice. On snow, the speed should be limited to 80 km/h or less. Even the red zone from the collision avoidance system should be extended up to 30 s and the yellow zone up to 45 s as can be seen in Table 4.5 (Figs. 4.6 and 4.7).

Besides the friction coefficient, other parameters influence braking distance [12, 15]:


- *Antilock Brake System (ABS)*. If the wheels are locked and sliding over the road surface, the braking force is a kinetic friction force only. Otherwise, if the wheels of the car continue to turn while braking, then static friction is operating. A test to stop the vehicle by locking the wheels and without locking the wheels showed that on a dry flat concrete ($\mu = 0.8$) with new tires the stopping distances were nearly the same. But the coefficient of kinetic friction is considerably less on a wet surface. As identified in [16] after analyzing 9 car models in 18 stopping situations, it was shown that ABS stops were shorter than those made without ABS with one exception: on loose gravel ABS stops increased with 27% overall. The same conclusion can also be found in the Finnish car magazine, "Tekniikan Maailma" and in the British car magazines "Evo" and "Autocar". The results

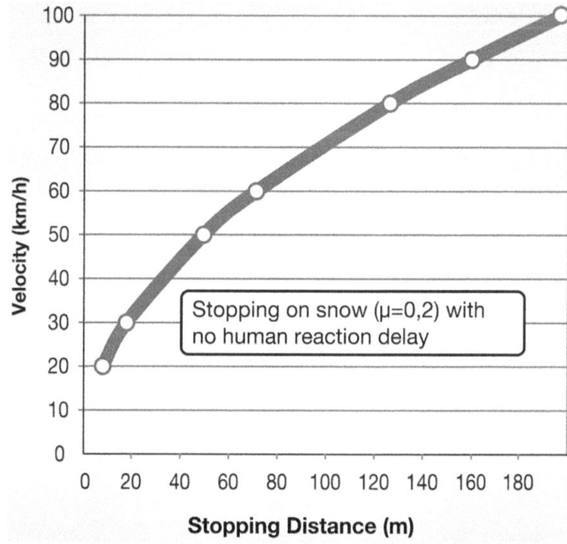


**Fig. 4.6** Stopping distance on snow ($\mu = 0.2$)

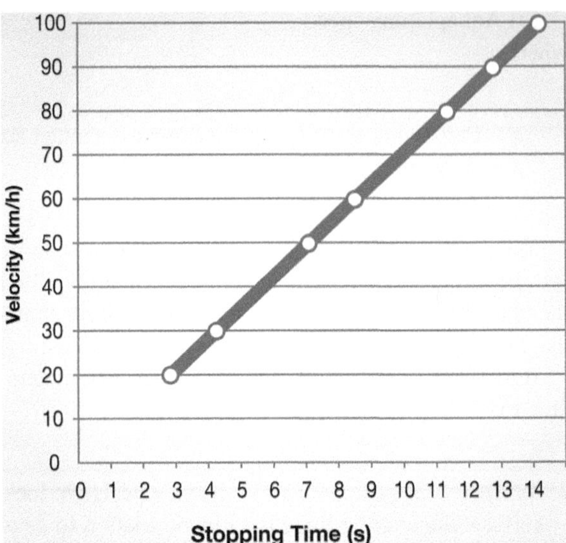



**Fig. 4.7** Stopping time on snow ($\mu = 0.2$)

demonstrate that on snow or gravel, locking the wheels reduces the braking distance! According to the Finnish car magazine, the stopping distance from 80 km/h on snow is 20% longer (on ice 58%) with ABS than without it.

- *Electronic Stability Program (ESP)* [17–19]. It compares the driver's intended direction in steering and braking inputs, to the vehicle's response, via lateral acceleration, rotation (yaw) and individual wheel speeds. ESP then brakes individually the front or rear wheel(s) and/or reduces excess engine power as needed to help correct understeering (plowing) or oversteering (fishtailing). As identified in [19] Continental Teves launched ESP II – the first Electronic Stability Control with steering intervention introducing an active steering control function (ASC), which makes a step further to a more controlling and precise braking force. ASC improves traction under acceleration by preventing the driving wheels from spinning on slippery surfaces. It also enhances vehicle stability by suppressing skidding in an emergency evasive maneuver or the result of other sudden steering inputs.
- *Tires condition/width/pressure.* According to [10] worn tires (to a tread depth of 1.6 mm) on a dry asphalt provides a better static friction coefficient (0.95 instead of 0.8) than the new ones. The situation is the other way around in wet/snow conditions.
- *Vehicle weight (loaded/unloaded).* According to Table 4.3 for cars at full load a maximum of 20% increase of stopping distance occurs (with one exception of 35%). The same increase value (35%) applies for trucks as well [20].
- *Straight/curved road.* As seen in Table 4.3, on curved dry asphalt road, the braking distance is as bad as on wet straight asphalt or even worse (increase of up to 60% in some isolated cases). In a curve lateral acceleration appears [21].

- *Brake reaction time.* A driver's recognition and reaction time may be from 1 to 2 s or even 3 s [12] opposite to the collision avoidance system that may have response times of milliseconds.
- *Car braking system* (discs on all wheels vs. on two wheels, hydraulic brake vs. electronic, etc.)
- *Suspension system* [21]
- *Wind speed/direction*

Therefore, there is a difference in the braking distance between different car manufacturers/models ($\pm18\%$ from theoretical value) [15] as seen in the following Table 4.1:

It should also be noted that the theoretical values of stopping distances are confirmed by real field tests as seen in Table 4.2:

As identified in [16, 20] the braking distances for nine different car models with ABS in different conditions are presented in Table 4.3.

At full load an increase in stopping distance generally occurs, especially for trucks [20].

Because it is hard to retrieve all of the many parameters that influence braking, I believe an error probability should be introduced: i.e. if the computed value of braking distance is x seconds, it should consider k × x, where k is a "confidence" factor that may take values from 1.1 to 1.3. Also a "back-up" system based on the

**Table 4.1**  Stopping distances on dry asphalt

| Car model | Distance (from 120 km/h) |
| --- | --- |
| Ferrari 550 Maranello | 59.7 m |
| Mercedes SLK230 Kompressor | 62.7 m |
| BMW Z3 (2.8) | 64.5 m |
| Saab 9000 Aero | 66.3 m |
| Porsche 911 Carrera 4 | 66.9 m |
| Nissan 200 SX | 68.4 m |
| Theoretical value (calculated above) | 70.9 m |
| Nissan maxima | 72.9 m |
| Lexus ES300 | 73.8 m |
| Honda Integra GS-R | 74.4 m |
| Mazda MX-5 | 76.8 m |
| Audi A4 | 80.7 m |
| Toyota Camry V6 | 82.2 m |
| Toyota Corolla | 95.7 m |

**Table 4.2**  Stopping distances

| | | |
| --- | --- | --- |
| Theoretical value (calculated above) | $\frac{31.5\,\mathrm{m}}{(\mu=0.8)}$ | $\frac{84\,\mathrm{m}}{(\mu=0.3)}$ |
| VW Golf V with non-studded tires (according to "Tekniikan Maailma" car magazine) | 32 m | 64 m |

**Table 4.3**  Stopping distances with ABS

| Car model | Stopping distance [m] | | | | | | | | | |
|---|---|---|---|---|---|---|---|---|---|---|
| Road geometry | Straight line | | | | | | | | Curve (91.4 m radius) | |
| Speed | From 97 km/h | | From 64 km/h | | From 80 km/h | | From 56 km/h | | From 80 km/h | |
| Surface type | Dry concrete | | Wet concrete | | Wet asphalt | | Loose gravel | | Dry asphalt | |
| Load | Light | Full | Light | Full | Light | Full | Light | Full | Light | Full |
| Year: 1995, 4 wheels speed sensors, 4 hydraulic channels | 42 | 48 | 23 | 25.5 | 30 | 33 | 32.5 | 30 | 33.5 | 46 |
| Year: 1994, 4 wheels speed sensors, 4 hydraulic channels | 44 | 53 | 23 | 25.5 | 31 | 36 | 30 | 27 | 35 | 37.5 |
| Year: 1997, 4 wheels speed sensors, 4 hydraulic channels | 42 | 48 | 25 | 27 | 31 | 35 | 30 | 30 | 35 | 35.5 |
| Year: 1996, 4 wheels speed sensors, 3 hydraulic channels | 44 | 49 | 25 | 27 | 34 | 34 | 28 | 31 | 36.5 | 38.5 |
| Year: 1996, 4 wheels speed sensors, 3 hydraulic channels | 43 | 43.5 | 32 | 30 | 35 | 36 | 32 | 33 | 35 | 37 |
| Year: 1997, 3 wheels speed sensors, 3 hydraulic channels | 44 | 48 | 25 | 25 | 32.5 | 36.5 | n/a | 34 | 53 | 59 |
| 4 × 2 Truck | 55 | 74 | n/a | n/a | n/a | n/a | n/a | n/a | n/a | n/a |
| 6 × 4 Truck | 55 | 74 | n/a | n/a | n/a | n/a | n/a | n/a | n/a | n/a |

last brake should be considered, i.e.: if the vehicle's last brake was used at "x"% intensity to reduce the velocity by "y" then the stopping distance (time) may be computed accordingly.

## 4.4  Time and Distance for Overtaking

The distance traveled during the overtaking maneuver is (Fig. 4.8):

$$s = v_0 \cdot t + \frac{1}{2} a \cdot t^2$$

(8) Distance traveled during overtaking
Where:

**Fig. 4.8** Vehicle distance
to overtake

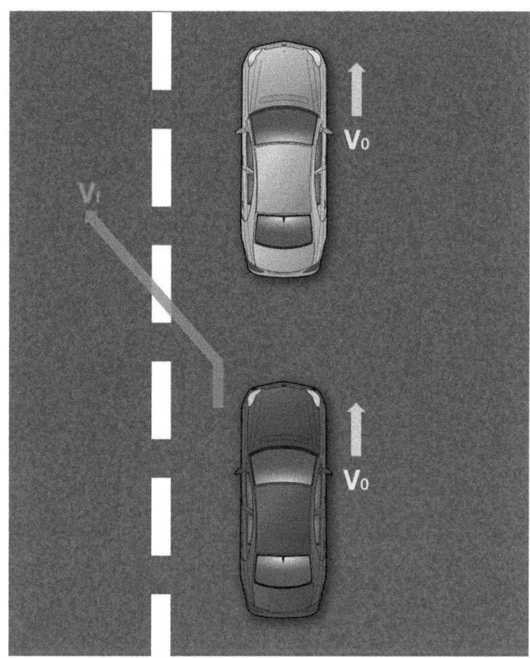


$a = \frac{v_f - v_o}{t}$ – medium acceleration (may be determined by accelerometers),
$v_o$ – initial velocity before overtaking maneuver,
$v_f$ – final velocity after overtaking maneuver,
$t$ – time to complete the acceleration.

The acceleration above is presumed constant, but in life scenarios it is variable. To predict this acceleration, the following may be used.

A database from previous tests (may be determined by accelerometers), manufacturer specifications. Table 4.4 is an example as identified in the car magazine

**Table 4.4**  Acceleration distances

| Time [s] | Velocity [km/h] | Acceleration [m/s$^2$] | Acceleration [g's] | Distance [m] | Cumulated distance [m] |
|---|---|---|---|---|---|
| 0 | 0 | 0 | 0 | 0 | 0 |
| 2.9 | 48 | 4.62 | 0.47 | 19.44 | 19.44 |
| 4.3 | 64 | 3.19 | 0.33 | 21.90 | 41.34 |
| 6.2 | 81 | 2.35 | 0.24 | 38.21 | 79.56 |
| 8.4 | 97 | 2.03 | 0.21 | 54.08 | 133.64 |
| 11.5 | 113 | 1.44 | 0.15 | 90.06 | 223.70 |
| 14.9 | 129 | 1.31 | 0.13 | 113.97 | 337.67 |
| 16.7 | 135 | 0.99 | 0.10 | 65.97 | 403.64 |
| 19.5 | 145 | 0.96 | 0.10 | 108.88 | 512.51 |

"Car and Driver", Nov. 2000. Only time and velocity are field measured values, the others are calculated (with above formulas).

## 4.5  Time Zones for Proactive Applications

We analyze time constrain for proactive applications. In future applications such as Cooperative Adaptive Cruise Control [22], Cooperative Forward Collision Warning, Lane Change Assistance and Pre-Crash Sensing will, together, form a Cooperative Collision Avoidance (CCAS) system. We define time zones in order to avoid crashes for such a system. Rear-end collisions all over the world cause a significant percentage of all accidents. This is why today we already have first "simple" applications such as safe following that provides assistance to the driver primarily to follow a vehicle ahead and to avoid rear-end collisions with other vehicles. But the system will eventually be enhanced to provide ultimate collision avoidance. Safe following, and of course the collision avoidance system, requires co-operative driving [23–25, 9] which means a complete reconstruction of the driving context and road environment using in combination on-board sensor data and cooperative system information transmitted over an ad-hoc vehicle network. In the first phase of introduction, when a vehicle makes a prediction of a collision, only warnings (visual, auditory, and/or haptic) are sent to the driver. Later, when reliability of those systems have been demonstrated, the system will take control (i.e. brake, steer) and override the human drivers.

A vehicle-to-vehicle communication system is required in order to obtain surrounding vehicle's dynamics. Each vehicle may be sender or receiver (Fig. 4.9). Thereby, the system must be able to send messages independently of receiving, otherwise big delays arise as outlined in Sect. 4.2.

The receiver has to obtain data information of the surrounding vehicles and the vehicle's logic has to evaluate the contents of the data information from surrounding vehicles. We summarize the data and network requirements of the collision avoidance system as well as possible devices/sensors that may be used in order to obtain the necessary information.

### *4.5.1 Data Requirements*

The DATA requirements of the collision avoidance system [26–28, 9] based on permanent beacon messages, but not limited to, are:

- *absolute position* [°] (by a localization service system such as GPS) and *relative position* [m] (by radar based sensors),
- *velocity* [m/s] (by speedometer, GPS),
- *acceleration/deceleration* – longitudinal and lateral [m/s$^2$] (by an XYZ accelerometer, g (Z-axis) must also be measured as it is not exactly the same all over the world),

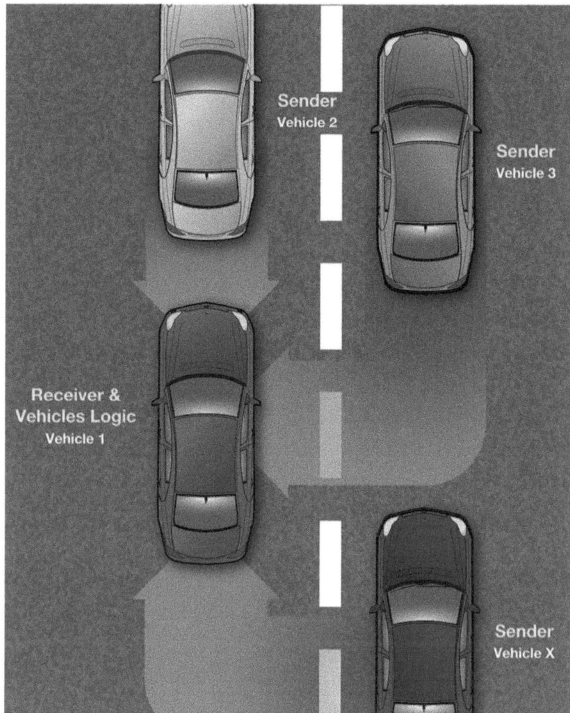


**Fig. 4.9** Decentralized message dissemination

- *heading* angle [deg] – yaw from the north (by GPS, compass, gyroscope),
- *yaw-rate* [deg/s] – longitudinal angular rotation (by GPS, gyroscope),
- *position confidence level* (based on the localization reliability – influenced by signal strength, weather conditions, etc.)
- *turn signal status* (left, right or no turn signal)
- *other vehicle's internal parameters*: ABS & ESP status, mass, suspension system, brake system, brake response time, tires pressure, width & condition (by on-board computer, sensors)
- *road friction coefficient* (by "intelligent" sensors outside: on the vehicle or in asphalt, center of traffic information),
- *wind speed and direction* (by anemometer, infrastructure communication)
- *road geometry (straight/curved)* (by GPS with digital map, radar "tracking")
- *current lane* (based on position and digital map)

Please note that the DATA requirements above refer to self vehicle as well as surrounding vehicles.

Other parameters that should be real-time monitored but only send if changed (critical situation: roll-over, pitch due to slippery in a curve for example).

- *pitch angle* [deg] – inclination of the front compared to the back of the vehicle (gyroscope)
- *pitch-rate* [deg/s] – lateral angular rotation (gyroscope),

- *roll angle* [deg] – sideways tilt of the vehicle (gyroscope)
- *roll-rate* [deg/s] – roll rotation (gyroscope),

The above parameters are interdependent. For example, roll-over may occur only in situations where the road friction coefficient is sufficient (i.e. dry asphalt) to resist sliding. Roll-over risk is computed on measurement of lateral acceleration and estimation of the gravity center of the vehicle. This depends on other factors such as vehicle mass, engine torque, tire size.

The collision avoidance system has to keep a very low rate of false alerts by using multi-sensor technologies such as: [29–31] vehicle communication, radar, GPS with digital map, camera vision. It also has to store information in an in-vehicle database, that may be used as "black-boxes" which, like their aircraft counterparts, record the last actions of the driver and vehicle prior to an impact [32, 33].

## *4.5.2 Network Requirements*

The NETWORK requirements of the collision avoidance system [1–3, 9] are:

- *Range of communication* (single-hop): between 10 and 300 m [34] (1,000 m with the new upcoming standard 802.11p);
- *Very good delivery rate* (packets received per packets sent): 0.99;
- *Low latency* (delay): 20÷100 ms;
- *Message frequency* (update rate): 5÷20 Hz (i.e.: 200÷500 ms);

The sender has to use geomulticast techniques [23, 3, 34, 9] to broadcast messages to all surrounding vehicles within their range area (e.g.: 1,000 m) at periodic rates. The range area (power) may be varied according to the speed of the sender and of the surrounding vehicles, but this requires that all vehicles are equipped.

- Single-hop transmission
- Reliable communication in a high mobile ad-hoc network (relative vehicles speed up to 200–300 km/h);
- Reliable bandwidth;
- Reliable security [35, 9].
- Recovery from faults
- Messages collision avoidance [36–38, 9]: each message must have a unique ID of the vehicle (e.g.: IPv6), maintaining privacy of the vehicle in the same time.

*Switch to two-way communication (Pre-Crash Sensing)* in case of conflicting solutions are generated or a collision is no longer unavoidable [8, 9]. The two (or more) vehicles engage a fast and reliable bidirectional unicast connection. If avoidance of collision is still possible, one of the vehicle has priority and the other vehicle must take other maneuvers. Otherwise, if collision is imminent, the unicast connection allows better use of actuators such as airbags or seat-belt pre-tensioners. A particular scenario of a conflicting solution is where one of the vehicles is incapable to act correctly to prevent the accident (e.g.: if one vehicle is unable to brake, the vehicle in front may accelerate and even change lane if possible).

### *4.5.3 The Cooperative Collision Avoidance System*

*The Cooperative Collision Avoidance System (CCAS)* should be based on distance/time, specifically on time to crash, similar to the TCAS (Traffic Alert and Collision Avoidance System) used by aircrafts as identified in [39, 38, 40]. Three time zones are defined below:

- Green zone: clear of collision
- Yellow zone: collision warning (advisory only)
- Red zone: shortest time to collision avoidance (action – if in mode II, see Fig. 4.10)

These time values are calculated according to each use case (parameters, e.g.: braking distance and obstacle type, e.g.: pedestrians). This means the red zone may, generally, be up to 15 s (based on mathematical formulas of stopping time – see Sect. 4.3) and yellow zone up to 22 s.

As a vehicle approaches the area where a traffic alert may be required, the beaconing rate may increase to twice per second opposite to extended ranges where the rate may be once every second.

Using a collision avoidance system, when reaching the "redzone", it may automatically engage "full stop" achieving deceleration of $\mu \cdot g = 9.8$ m/s$^2$ (if $\mu = 1$) without loss of control [39, 38, 12]. Also, a driver's reaction time is about 2 s opposite to the collision avoidance system that may have response times of milliseconds (a 2 s delay at a speed of 120 km/h means a plus of 66 m of the stopping distance).

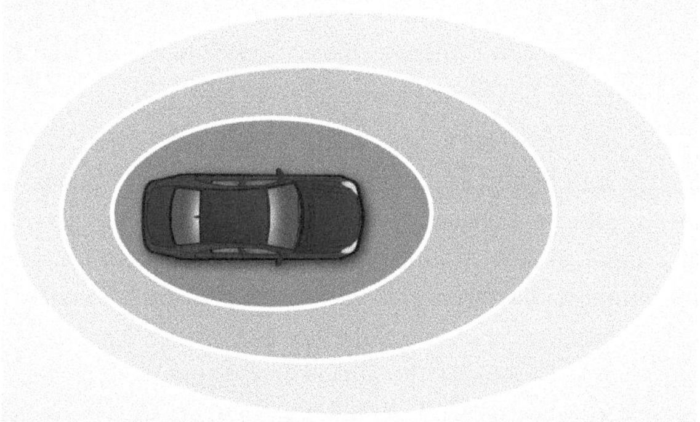

**Fig. 4.10** Safety zones

The heart of the collision avoidance system is the system logic. The logic requires a trade-off between necessary protection and unnecessary advisories. This trade-off is accomplished by controlling the sensitivity level (SL), which controls the time or tau ($\tau$) thresholds for the red and yellow zones, and therefore the dimensions of the protected space around each vehicle. However, as the amount of protected space increases, the incidence of unnecessary alerts has the potential to increase. Therefore, different sensitivity levels should exist according to the use case.

An example of this idea is presented in the Table 4.5 with only 9 sensitivity levels. Probably more steps for speed and $\mu$ should be taken into account and thus more SLs. Also, the $\tau$ values are presented for stopped obstacles (rear-end) or vehicles that come towards each other. In this second case, where it was assumed that both vehicles heading on each other with the same speed, the time is doubled. It was seen that the red zone, on snow, may grow up to 42 s! In real life scenarios, both vehicles' speed (head-on or rear-end) should be computed and then the time should be set accordingly.

All the values from the Table 4.5 are theoretical, calculated using mathematical formulas presented above and thus neglecting vehicle mass, tires condition, suspension system, response time, braking system, etc.

The collision avoidance system should have two operating modes:

– Alert only (I),
– Alert and Act (II). In certain special cases the mode I may be switched to II automatically by infrastructure, if required.

*The collision avoidance system needs to detect the road geometry, communicating objects, and non-communicating objects.*

The road geometry ahead [41, 42, 29, 30] (straight, curve, etc.) can be detected by using

- a digital map and GPS receiver to enable an indication of vehicle position and direction of travel on the map.
- a camera vision with integrated signal processing and imaging electronics to "follow" lane markings. This is not a reliable way to follow lanes because the lane may be missing or due to weather conditions not be visible.
- radar tracking to follow the vehicles ahead by monitoring their trajectories.

Communicating objects (equipped vehicles). These "intelligent" vehicles have to use antennas [38] in order to exchange messages between them. To transmit interrogations and to receive replies, directional as well as omnidirectional antennas may be used. A solution to minimize multipath interference and to proper adjust the strength of the signal is to use multiple parallel antennas. For example, two parallel antennas (one on top and one on bottom) may be used. The collision avoidance system includes the surveillance function that decodes the beacons from other vehicles. Potential threats are determined based on the surveillance function. This

**Table 4.5** Sensitivity levels [12]

| Relative position of the sides involved | Obstacle type | Friction coefficient ($\mu$) | Speed [km/h] | Sensitivity level (SL) | Time ($\tau$) [s] | |
|---|---|---|---|---|---|---|
| | | | | | Red | Yellow |
| Rear-end (stopped) | Car/Lorry/Motorcycle/Moped | >0.75 | <60 | 1 | 3 | 4.5 |
| | | | 60÷120 | 3 | 5 | 7.5 |
| | | | >120 | 4 | 7 | 10.5 |
| | | 0.45–0.75 | <60 | 2 | 4 | 6 |
| | | | 60÷120 | 4 | 7 | 10.5 |
| | | | 120÷160 | 5 | 9 | 13.5 |
| | Tree/Pillar | <0.45 | <60 | 5 | 9 | 13.5 |
| | | | 60÷100 | 7 | 14 | 21 |
| | | >0.75 | <60 | 2 | 4 | 6 |
| | | | 60÷120 | 4 | 7 | 10.5 |
| | | | >120 | 5 | 9 | 13.5 |
| | Pedestrian/Bicycle | 0.45 ÷ 0.75 | <60 | 3 | 5 | 7.5 |
| | | | 60÷120 | 5 | 9 | 13.5 |
| | | | 120÷160 | 6 | 12 | 18 |
| | | <0.45 | <60 | 6 | 12 | 18 |
| | | | 60÷100 | 8 | 18 | 27 |
| | | >0.75 | <60 | 3 | 5 | 7.5 |
| | | | 60÷120 | 5 | 9 | 13.5 |
| | | | >120 | 7 | 14 | 21 |
| Head-on | Car/Lorry/Motorcycle/Moped | 0.45 ÷ 0.75 | <60 | 4 | 7 | 10.5 |
| | | | 60÷120 | 7 | 14 | 21 |
| | | | 120÷160 | 8 | 18 | 27 |
| | | <0.45 | <60 | 8 | 18 | 27 |
| | | | 60÷100 | 9 | 28 | 42 |

function has as input parameters such as: range, position, speed, acceleration and bearing of nearby vehicles. The trick is to decode multiple overlapping beacons. Currently, in aircraft industry, are used some hardware degarblers to decode up to three overlapping replies.

For non-communicating objects, an accuracy level at object detection should be introduced. It can be detected by Cameras or range sensors described in the following paragraphs:

A camera vision [42, 29] with integrated signal processing and imaging electronics to detect the object: its edges, dimensions, position, color. Also, optical character recognition (OCR) can be achieved. The principle of this technology is to interpret an image. First, the best mathematical polynomial model that describes the image motion in a specified zone is computed. Then, by comparing the current image motion with previously computed motion (from a database) obstacles can be recognized. Third, the object trajectory as well as time to collision are computed and used as input for the surveillance function.

RADAR (ultrasound sensors), LASER and LIDAR (IR) [29, 43] to detect presence, absence, azimuth or distance of target objects as well as relative speed between the host vehicle and the vehicle or object ahead of it. The LASER and LIDAR provide less interference and false alarms than RADAR, but also poor performance if not clear (dirt, bad weather, etc.). To eliminate false alarms sophisticated signal processing and tracking algorithms need to be used. Generally, two types of Radars are used: short distance (approx. 30 m) with wide angle (70–90°) and long distance (approx. 1–150 m) with narrow angle (9–12°).

An example of multi-sensor communicating and non-communicating object detection in presented in Fig. 4.11. It can be seen that by wireless communication a complete driver situation awareness may be accomplished where normal radar field is not covered.

The intelligent vehicle system must be able to: [29–31, 9].

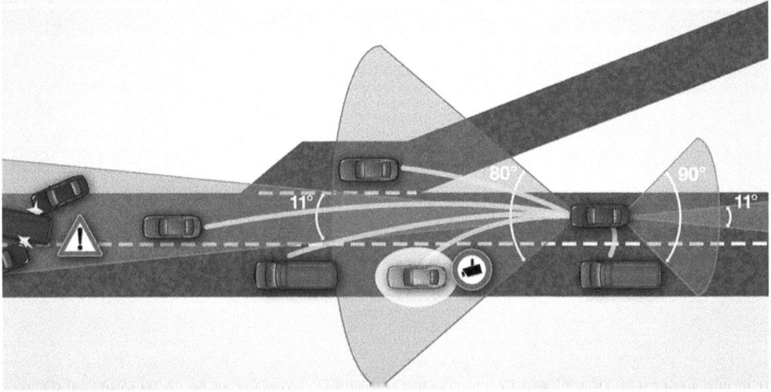


**Fig. 4.11** Example of complementary information from multiple sensors

- Synchronize data from different sensors by shifting the location data (e.g.: using some form of Kalman filtering);
- Collect the position and velocity of the object, shape of the object, road geometry from the multiple sensors and eliminate false warnings. The sense mechanism must be done in "real-time" to maintain a tracking of the surrounding objects;
- Analyze the information (e.g.: relative speed, shape, inside/outside of lane, etc.) in order to classify objects in: cars, bus, trucks, motorcycles, bicycles, pedestrians, animals, trees, road signs, etc.;
- Issue a resolution of the conflict.

We analyzed a system that creates a reconstruction of the driving context and road environment using a combination of vehicle communication and on-board sensor data.

## References


1. Q. Xu, R. Sengupta, and D. Jiang, Design and analysis of highway safety communication protocol in 5.9 GHz dedicated short range communication spectrum, USA
2. S. Yousefi, A. Benslimane, and M. Fathy, Performance of beacon safety message dissemination in vehicular ad hoc network
3. X. Ma and X. Chen, Performance analysis and enhancement of safety applications in DSRC vehicular ad hoc networks, 2007
4. H. Hada and H. Sunahara, DGPS and RTK positioning using the internet, Japan, 2000
5. http://www.wsv.de/fvt/index.html
6. http://www.gpscontrol.com
7. http://www.trimble.com
8. http://www.prevent-ip.org
9. Car2Car Communication Consortium – Application Working Group, 2006, www.car-to-car.org
10. http://www.tuninglinx.com/html/aquaplaning.html
11. W. Seibert and H.J. Hilt, WILLWARN Hazard detection, Böblingen (Germany), 15, November 2006
12. http://www.csgnetwork.com/stopdistcalc.html
13. R. Serway and J. Jewett, Physics for scientists and engineers, 6th ed., Brooks/Cole, 2004
14. B.N.J. Persson, U. Tartaglino, O. Albohr, and E. Tosatti, Sealing is at the origin of rubber slipping on wet roads, 2004
15. http://www.sdt.com.au/STOPPINGDISTANCE.htm
16. G.J. Forkenbrock, M. Flick, and W.R. Garrot, NHTSA light vehicle antilock brake system research program task 4: a test track study of light vehicle ABS performance over a broad range of surfaces and maneuvers, January 1999
17. http://www.bosch-esperience.com
18. http://www.nhtsa.dot.gov
19. http://www.conti-online.com/generator/www/us/en/continentalteves/continentalteves/general/home/index_en.html
20. R. Radlinski, Electronically controlled braking systems performance evaluation, February 2003
21. R. van der Aalst, Formula student vehicle analysis by means of simulation, 2005

22. A. Vahidi and A. Eskandarian, Research advances in intelligent collision avoidance and adaptive cruise control. IEEE Transactions on Intelligent Transportation Systems, Vol. 4, No. 3, pp. 143–153, September 2003
23. T. Kosch, Technical concept and prerequisites of car-to-car communication, München: BMW group research and technology, http://www.car-to-car.org
24. M. Provera, Activities and applications of the vehicle to vehicle and vehicle to infrastructure communication to enhance road safety, Torino, 2005
25. C. Adler and M. Strassberger, Putting together the pieces – a comprehensive view on cooperative local danger warning, München
26. S.J. Hong, J.Y. Choi, Y.I. Jeong, K.Y. Jeong, M.H. Lee, K.T. Park, K.S. Yoon, and N.S. Hur, Lateral control of autonomous vehicle by yaw rate feedback, Korea, 2001
27. http://www.oxts.com
28. http://www-nrd.nhtsa.dot.gov/
29. http://www.carsense.org/
30. Ivsource.net, Advanced Collision Avoidance (ACAS) Field test launched in Michigan, 2003
31. http://www.its.dot.gov
32. http://www.howstuffworks.com/black-box10.htm
33. http://www.l-3ar.com
34. DSRC_Tutorial_06-10-021.ppt
35. http://www.sevecom.org
36. D. Marsh, CANbus networks break into mainstream use, Europe: EDN, 2002, http://www.edn.com
37. R. Garces and J.J. Garcia-Luna-Aceves, Collision avoidance and resolution multiple access with transmission queues, California, 1999
38. U.S. Department of Transportation, Introduction to TCAS II version 7, November 2000
39. Ed Williams, Airborne collision avoidance system, Australia, 2004
40. R.Y. Gazit and J.D. Powell, Aircraft collision avoidance based on Gps position broadcasts, Stanford University, 1996
41. G. Segarra, Activities and applications of the car 2 car communication: the Renault vision, http://www.car-to-car.org
42. http://www.globalspec.com/
43. A. Hoess, Multifunctional automotive radar network (RadarNet) final report, RadarNet consortium, 2004

# Chapter 5
# Routing

Routing refers to move a data packet from source to destination and if required the assignment of a path to the destination. In multi-hop regime routing means to forward packets that contain information through other vehicles [1]. This information refers to alerts about events that already happened, like local danger warnings and traffic flow information. If no vehicle is within the communication range a packet is stored and forwarded as soon as a new vehicle comes into reach.

In multi-hop regime the information may be disseminated in two ways: to all surrounding nodes (multicast) or to the ones in a specific area (geocast). Nodes, by exchanging information about network links, compute the best path by which to route messages to other nodes.

Routing in highly mobile ad-hoc networks has to preserve the integrity of message information disseminated in the network while minimizing the number of propagations of each message. Different factors influence the message integrity, e.g. routing algorithm, environmental conditions (physical layer), but also intruder attacks (security). Several algorithms are presented below, and in order to compare them we introduce three vital parameters: PDR, latency and overhead.

The PDR (Packet Delivery Ratio) [2, 3] represents the successfully received packets per sent packets. In theory, if no loss of messages occurs, this value should be 1. In real tests this value is below 1 because in ad-hoc networks one big weakness is that route between a source and a destination is likely to have errors or even break during communication. It was demonstrated in [4] that the data packet loss rate can be decreased significantly by using relative local positions between nodes to discover routes and to make the routing decision for the ad-hoc network.

The latency or end-to-end delay [5] represents the time delay from the source sending a packet to the destination receiving it. RTT (Round Trip Time) represents the latency plus time delay from destination back to source, also known as "ping".

The routing overhead, which represents ratio of control data per payload data should be as low as possible to prevent excessive load of the network. The overhead can be reduced using location-based routing algorithms that may obtain absolute position from a positioning system such as Global Positioning System (GPS) or relative local position between nodes.

Other important parameter in routing algorithms is the fault tolerance [6] that requires detection of and recovery from faults.

R. Popescu-Zeletin et al., *Vehicular-2-X Communication*,
DOI 10.1007/978-3-540-77143-2_5, © Springer-Verlag Berlin Heidelberg 2010

The above parameters can be unified under the so-called scalability effect. This effect [7–9] ensures adaptability and represents QoS (Quality of Service) impact (PDR, latency, overhead, fault tolerance, etc.) with the growth of network size. A scalable algorithm maintains its QoS parameters with the increase of the network.

Routing with multi-hop position based regime is illustrated in Fig. 5.1. Different transmission strategies are possible: Either broadcast the message to all surrounding vehicles until the geocast region is reached, or create a path from source to the geocast region. Each transmission has its advantages and disadvantages. If a path is created, than less network load is required, but if the link is broken, information misses its target. If the flooding approach is used, then multiple paths exists. So, fault tolerance is provided, but also an increase of network load.

A routing algorithm is composed of two basic components: A data distribution algorithm and a route decision algorithm. A router typically obtains routing information via its data distribution algorithm, and it uses this information to construct a forwarding table by applying the protocol's decision algorithm to the received routing data. The distribution algorithm can be on-demand or fully distributed. On-demand (reactive) means that the path routes are computed only when specific nodes request this, while fully distributed (proactive) means that the path routes to all possible destinations are already known. The proactive approach provides shorter latency, but routes may not always be updated (slow adapt to change). So, a lot of control packets are required resulting in an increase of overhead. On the other hand, the reactive approach provides less overhead but longer latency. Therefore, the best approach, called hybrid, combines the advantages of both previous approaches.

Next, we present different forward routing algorithms, beginning with non position based ones followed by position based ones and ending with some particular algorithms that provide security within the routing layer. Based on the analyzed algorithms, we find CLA and Octopus with AGF technology on top to

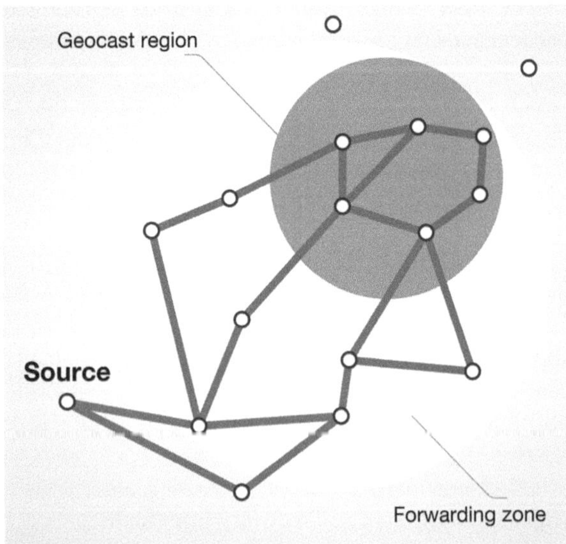


**Fig. 5.1** Multi-hop position based regime [11]

provide the best promising results (summarizing some features: low load, good PDR, fault-tolerance and hybrid approach). Unfortunately, we cannot conclude a best algorithm due to missing of some actual value parameters and no real field implementation.

On top level, two large groups of routing protocols are presented: first common routing protocols, such as AODV, and second a group of secure routing protocols and security extensions to existing routing protocols. Focus is on routing protocols that use locations to support routing functions; this is because these protocols are assumed to behave best in the highly mobile environment of vehicular communications.

## 5.1  Multi-hop Routing Protocols

The following sections present common routing protocols with focus on position based routing protocols.

### 5.1.1  Ad Hoc on Demand Distance Vector (AODV)

A well known routing algorithm designed for mobile ad hoc networks is the Ad hoc On Demand Distance Vector (AODV) [8, 9]. AODV is described in detail in RFC 3561 [10]. AODV supports both unicast and multicast communications. Routes are created on demand and maintained as long as they are needed by the sources. The management of multicast group members are is done in a tree structure. The tree structure reflects the network topology necessary to connect the members of the multicast group. Nodes and edges represent network nodes and connections, respectively. Sequence numbers are used to ensure that routes are up to date. Main properties of AODV are its ability to prevent loops and its scalability even for a large number of nodes.

Core mechanisms for route construction in AODV are the route request (RREQ) and route reply (RREP) (see Fig. 5.2) [8, 9, 12].

Before initiating a transmission to an unknown node, the source of the transmission broadcasts a RREQ packet. Upon reception of the RREQ, receivers update their local topology information, i.e., the routing table, and keep backwards pointers to mark the source of the RREQ, this is called the *Reverse Path*.

The RREQ is identified by its the RREQ ID and its source node IP address It contains sequence numbers for originator and destination; the originator sequence number is used to for the backward reference in the routing tables of the receiving

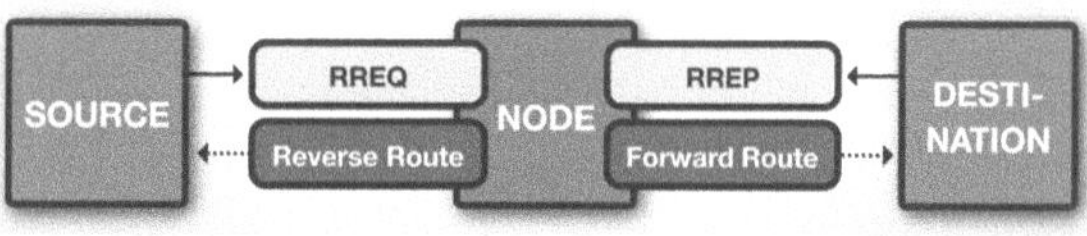



**Fig. 5.2** AODV route creation using route reply and route request primitives

nodes. The destination sequence number is the last sequence number the source received for a route to the destination.

A node generates a RREP, if it either is the destination, or its routing table contains an entry for an active route to the destination with a destination sequence number greater or equal than that in the RREQ. If no RREP is generated, the RREQ is rebroadcast with updated fields for hop count and for the destination sequence number (to a potentially newer entry in the forwarding node).

When the RREP is passed backwards through the nodes on the path to the source, they create an entry in the routing table to the destination node (known as *Forward Path*), before propagating the RREP. After reception of a RREP to its request, a node starts to transmit data packets to the corresponding destination (through the just established route). RREP that arrive later are only used to update the source's routing table, if the incoming route is better, e.g. has a lower hop count.

When the source stops sending data packets, the links will time out and eventually be deleted from the intermediate node routing tables. In case the destination(s) are not to be found, an error message (RERR) is transmitted back to the source. So, the discontinuity of the link is announced.

In order to create multicast routes the principle remains the same: broadcast of a route request (RREQ). But this time the broadcast is sent to the IP address of the multicast group with the "J" (join) flag. The node(s), from the multicast tree, that has the most recent sequence number respond to the RREQ with a RREP. Backwards pointers are set in the process to keep the multicast route tables.

After expiry of the specified discovery period, the route is activated using a Multicast Activation (MACT) message unicast by the source noted to its selected next hop.

The routes, including the multicast tree, are kept as long as they are active. During the lifetime of a route, link breakages will occur because of the nodes' mobility.

To better understand the working principle of AODV we will illustrate a concrete example in Fig. 5.3. Here, we present the links kept by the nodes for a particular situation. First, we presume that the route is created from source (S) to destination (D) via nodes: 6, 7 and 5 according to AODV algorithm. This is called the forward link (RREQ). The destination node (D) responds to the source node (S) and creates

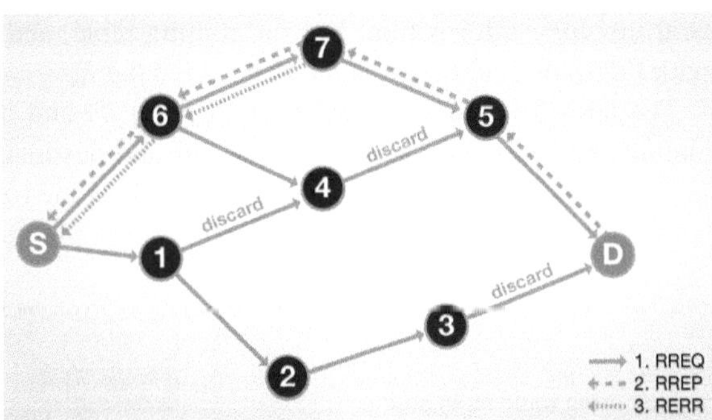


**Fig. 5.3** AODV routing example [13]

the reverse link (RREP). But, if after a while, node 5 moves away from node 7, a link failure occurs. Then, node 7 detects link failure and broadcasts RERR back to source.

Performance is seriously affected [13] (huge overheads & data transfer interruptions) if route re-discovery process is too frequent due to node mobility or bad channel conditions.

As also identified in [14], in VANET, especially in city scenarios, AODV performance problems occur. These problems are collisions that occur due to redundant transmission (uncontrolled flooding) and sometimes lead to lost transmissions and prevent the dissemination of information to all the nodes. From the experiments in the city scenarios the authors of [14] conclude that half of the RREQ (route requests) are dropped by nodes, due to collisions.

As identified after studding many different simulations [15, 2, 13, 16–24]. AODV provides long latency time, high overhead and low PDR in mobile environment. The algorithm is not position-based (topology based) and it does not provide fault tolerance.

Caching may be introduced [25] to improve channel load and often reduce the number of hops.

## 5.1.2 Grid Location Service (GLS)

GLS proposed by J. Li et al. in [26] as discovery service is based on the idea that each node sends its current position only to a small subset of all nodes.

For multi-hop routing all vehicle-to-vehicle nodes have to be able to determine their position [1]. Based on the position, the closest node to the destination is chosen to forward a packet (greedy technique). So the position is vital to the greedy forwarding algorithm used to avoid flooding of the network. But in real vehicle-to-vehicle networks, especially in cities where obstructions exist, the pure greedy technique could have a bad consequence: lost packets. This happens when there is no neighbor available that is closer to the destination than the current forwarding hop [15].

In GLS each node in the network maintains only a part of the global location database [27]. A collection of neighbor nodes forms a cell. The service creates a fixed grid of a location database, whose grid lines are known to all nodes in the network. Each node has a unique identification number (ID) that consist of a natural number. Neighbor nodes are grouped in a cell and each node keeps tracking of the neighbor nodes within the same cell. Based on the transmission range of the nodes, one and two hop neighbor lists are created.

A node that stores the mapping of another node ID to its location is named location server. The location of one node does not refer to the actual geographical position, but a quantized location in the network database. So, to reach the "location" of one node is enough to locate the corresponding cell. The location server is required to find communication partners and needs to be updated when a node changes its position.

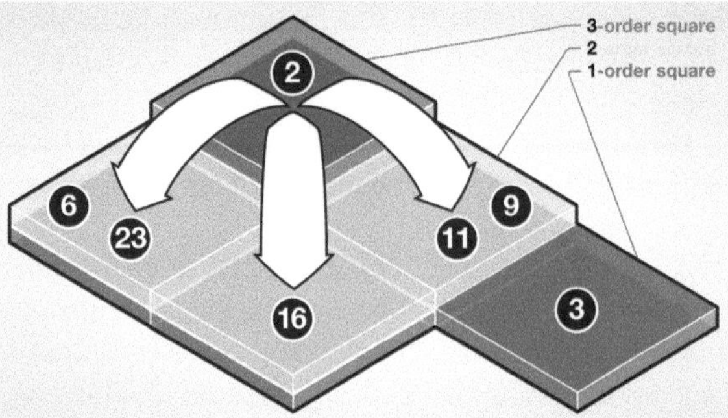



**Fig. 5.4:** GLS grid overview [28]

The basic overview of GLS is to create the database in the form of a quadtree on top of the grid structure (Fig. 5.4). A quadtree is a tree data structure in which each internal node has up to four children.

As depicted in Fig. 5.4 the grid cells are organized in layers, where the original bottom grid cells are called 1-order squares, followed by 2-order squares, and so on. The basic principle of GLS is to select one location server in the n-order square for an arbitrary node "A" in n+1-order square.

To determine the square for the location server (for node A) a hash function is used. The hash function takes as parameters the ID of node A as well as all the IDs of the nodes in the respective square.

The location server for node A in an n-order square is the one that has the "nearest" ID to node A's ID in the respective n-order square. The hash function is designed so that each node can use it for a specific square and target ID in order to find the location servers of the target ID.

The algorithm used to find a location server for the mapping of an arbitrary node ID to that node's location is next presented. Each of the square keeps a track from the node IDs to the current location server. To locate a node the principle is to find the node with the nearest ID the querying node knows of. The node with the "nearest" ID knows a node with a nearer node ID and so on. The process continues until a node that has the queried location information is found. When the target node is found, it will send back a response to the query originator node to indicate its location.

The location update process resembles the algorithm explained above of finding a node, namely: the location update messages are sent to n-order squares and then forwarded to nodes with nearer IDs within a square. The difference from the above algorithm is that the information is written as opposed to read.

An example of a location server/update process [23, 29] is presented in Fig. 5.5. First, each node knows its neighbor in the same cell: 1-order square. Then, node 2 sends location updates (red arrow) to its servers (9, 16, 6) – 2-order square. Node 3 sends location updates to its servers (2) – 3-order square. When node 23 sends a query (blue arrow) for 3 it begins to search for node 3 from "hop" to "hop" (it doesn't directly know 3, ask its neighbor which knows 2, then 2 knows 3).

One important issue [27] is to minimize the time when links between nodes have outdated information. This happens when a node moves into a new square, resulting that some of the other nodes remain without location server.

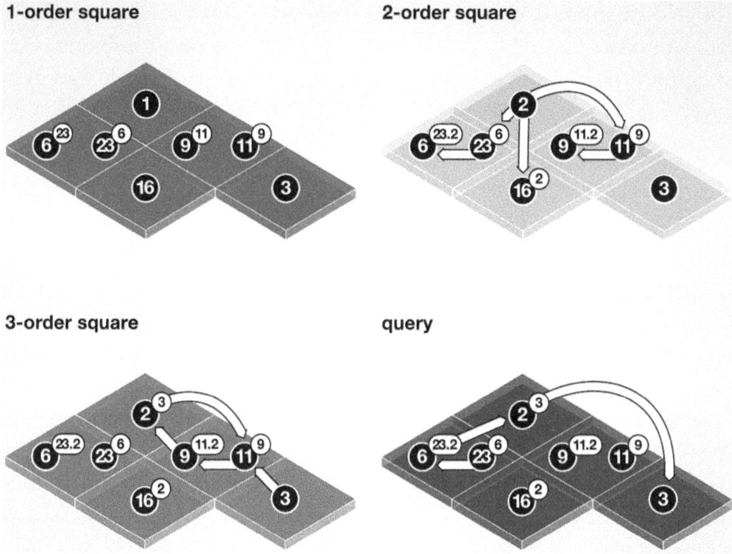



**Fig. 5.5** GLS process example [29]

Thus, a specific correlation between bandwidth consumption of GLS signaling messages (in particular, location updates) and the success rate of location queries exists. Therefore a trigger for sending a location update is used. After a movement of:

$$2^{n-2} \cdot d$$

(9) Location update trigger
a new update of location server is required. Where:

n – is the n-th order location server;
d – distance a node might travel until a location update is sent again.

Still, drawbacks exists. It is possible before update process is performed some nodes remain without location server which:

- may lead to packet loss (slow update);
- due to outdated neighbor table entries excessive re-sending of data may occur (overload);
- if update is too often results in excessive overhead.

After studying many different simulations [26, 20, 23, 30, 31, 28, 32–34], GLS provides long latency time (though shorter than AODV), high overhead and low PDR in mobile environment. Latency and overhead may increase up to 3 times with node density. The algorithm provides a fault tolerance: more than one location servers in grid ensure that even if a single location server for node n fails, n won't be unreachable. As opposed to AODV algorithm, GLS is position-based and uses a proactive approach resulting in low privacy. It is considered scalable, but in big networks (over 1,000 nodes) PDR drops below 0.5. It eliminates uncontrolled flooding issue, present at AODV, by using geographical dissemination.

### 5.1.3 Greedy Perimeter Stateless Routing (GPSR)

Greedy Perimeter Stateless Routing (GPSR) is an algorithm that consists of two methods for forwarding packets [25]: *greedy forwarding*, which is used wherever possible, and *perimeter forwarding*, which is used in the regions where greedy forwarding cannot be.

The greedy forwarding algorithm uses packets that carry the locations of their destinations. The packets are stamped by the source node. This way, the packets are always forwarded to the neighbor that is geographically closest to the destination.

The drawbacks of pure greedy forwarding [7]:

- The position accuracy drops if the nodes move (mobility). It is possible that a location server node changes its position and before update process is performed some nodes remain without location server. This may lead to packet loss. Also, due to outdated neighbor table entries excessive re-sending of data may occur.
- Additional network load due to the beacons
- Missing of recovery from failure due to the link-layer broadcast of the beacons. This leads to failure in transmission, because nodes being close to each other are not recognized as such.

The recovery strategy of the GPSR called *Perimeter Mode* [15, 25] is used in order to avoid the lost packets that may occur in pure greedy technique when there is no neighbor available that is closer to the destination than the current forwarding hop. The perimeter mode of GPSR consists of two elements. First, a distributed planarization algorithm that locally transfers the connectivity graph into a planar graph by the removal of "redundant" edges. Second, an online routing algorithm for planar graphs that forwards a packet along the faces of the planar graph towards the destination node.

The above algorithm still has a series of drawbacks in a city scenario, such as:

- Network interruption in some particular cases due to wrong elimination of "redundant" edges;
- Too many hops (compared to pure greedy technique);
- Routing loops (in some particular cases, the "Right Hand Rule" that is used creates these loops).

In GPSR, when packets are forwarded in greedy mode [14], the selection of the next-hop node is based on the neighbor's table of the forwarding node. But the neighbor's table is not always up-to-date, so the selected neighbor may not be the best choice or even not be a neighbor any more.

Based on different simulations [20, 21, 24, 32, 25], the channel load is low on average, but packet cost increases quickly as node density decreases. Average latency for GPSR is virtually unrelated to packet rate, but increases with network size. It provides a good PDR but it lowers in mobile environment. The algorithm is position-based, fault-tolerant and uses a proactive approach. It eliminates uncontrolled flooding issue by using geographical dissemination.

### *5.1.4 Geographic Source Routing (GSR)*

The Geographic Source Routing (GSR) is a position-based routing method [15] that takes into account a map of an area (navigation system with GPS is required) and that eliminates the drawbacks of the GPSR in a city scenario. To find the current position of a desired communication partner the Reactive Location Service (RLS) is used.

RLS is a route discovery procedure that allows a mobile node in an ad hoc network to inquire the geographic position of another node in an on-demand fashion. As opposed to a proactive location service (such as GLS), where location information is distributed and managed even when there is no node in the network requesting this information at that point in time, a reactive location service only retrieves location information "on-demand".

The sending node computes a path to the destination. The path is actually a sequence of junctions based on a digital map of streets & obstacles such as buildings so that the transmission may bypass the obstacles.

The principle of RLS [35] is next depicted. When a node wants to reach another node, it issues a query, which contains the source node ID and location, as well as destination node ID. To reach the destination node, the flooding technique is used. If the destination is not found within a certain time window, the flooding is discarded. The time window is known as time-to-live (TTL). The destination node (if reached) replies with a packet back to the source. The reply packet contains the IDs and locations of both the source (now as destination) and destination nodes. The location discovery cycle is completed if the reply packet is received by the source node.

The source node stamps all the query packets with a number (sequence number). The number represents the number of attempts of the source node to reach the destination node's location. This sequence number in stored in a cache memory and it is used to avoid infinite packet looping and duplication when the network is flooded with information. The principle is that each forwarding node compares its sequence number with the one in cache and, if it is smaller than that, the query is discarded due to duplicated or looped packet.

With the following information [15]: location query packet and location reply packet, the sending node can calculate a path to the destination based on the map of the environment (e.g.: streets and buildings). The transmission of packets is done through pure greedy technique only when no block (obstacle) is found between two successive crossings. Otherwise, in order to reach the destination, the sending node computes a sequence of paths (junctions) to be followed.

The sequence can be put into the packet header or can be computed by each forwarding node. Each solution has advantages and disadvantages:

- Using header variant eliminates power computation required by each forwarding node but no recompute of the route is possible;
- Using "the computed by each forwarding node" variant requires more computing power but a node may recompute another route to the destination based on the information from its neighbors.

As a drawback for GSR: requires enough vehicles on a street to provide connectivity between two involved junctions. Other path-finding strategies may be used, such as infrastructure. The infrastructure is required in order to forward the packets if no other vehicles are on the street.

### 5.1.5 Contention-Based Forwarding (CBF)

Contention-Based Forwarding (CBF) [36] is a mechanism for position-based unicast forwarding, without the use of neighborhood knowledge. Instead, all suitable neighbors of the forwarding node participate in the next hop selection process and the forwarding decision is based on the actual position of the nodes at the time a packet is forwarded. This algorithm eliminates the drawbacks of pure greedy solution.

In position based routing [7] the principle is that the forwarding of the packet, from one hop to another, is done based on the local geographical position of the nodes. Being based on local position information of each node, it is not necessary to create and maintain a global route. Therefore, the algorithm is generally highly scalable and robust against network mobility.

The CBF mechanism [36] uses a contention-based algorithm to determine the next node forwarder and to keep silent the other nodes. Normally, CBF supports unicast routing, but can be used in VANET's with information dissemination, so that the packet would be disseminated in several directions at the same time.

Its main advantages are:

- all relevant nodes are involved in the decision making, i.e.: decision is based on the current position of all neighboring nodes;
- low overhead, high scalability and high adaptability to the network mobility due to the missing of neighborhood table or knowledge as well as beacons.

The usage of message content is relatively new. This brings new concepts such as self-organizing dissemination area for a specific lifetime as well as message prioritization (establishment of messages relevance). The relevance of the message is based on context, such as:

- distance/time to the event;
- distance/time to the last forwarder of the message;
- probability that the information is already known by the neighbor vehicles;
- time since the generation of the message;
- time since the last successful transmission;
- time since the last successful reception of the message;
- heading (driving direction);
- power transmission of a vehicle.

The basic principles after which this scheme functions [36]:

- The forwarding vehicle broadcasts the packet to all surrounding neighbors;
- One "best of" neighbor is selected to forward the packet;
- Then, the "best of" neighbor suppresses the other nodes to avoid packet duplication.

The CBF technique is based on timers. This means the packet may not be directly forwarded by a node, but kept for a certain period and then transmitted. The packet is forwarded by the node closest to the destination (whose timer expires first). It is possible that more than one node might forward the message due to the propagation delay, but the chances lower as the node is closer to the destination.

The contention can be realized on the basis of biased timers and three alternative suppression strategies [7]:

The Timer-Based Contention is further explained below. The packet progress $P$ (normalized to the radio range) is calculated with the formula:

$$P(f, z, n) = \max\left\{0, \frac{\text{dist}(f, z) - \text{dist}(n, z)}{r_{\text{radio}}}\right\}$$

(10) Packet progress
where:

    f represents the position of the last forwarder that can be read out of the packet header;

    z is the destination of the packet;

    n is the position of the node that is contending for the right to forward the packet;

    dist() function calculates the Euclidean distance between two positions (e.g. GPS positions);

    $r_{\text{radio}}$ stands for the maximum radio range.

To further minimize the remaining distance to the destination of the packet, each node calculates its timer based on this progress function. The timer t of the nodes is set as:

$$t(P) = T(1 - P)$$

(11) Timer

Where:

T is the maximum forwarding delay.

The node in the position n resents the packet when the timer, t(P), expires. The contention process is based on this timer (earliest forwarder wins). Then the timer is discarded to eliminate duplicate rebroadcasts.

It is possible that more than one node might forward the message due to:

- the propagation delay (can be avoided increasing the maximum forwarding delay T)
- the out of range of each others radio (in the worst case, up to three copies of the packets may be forwarded, as shown in Fig. 5.6)

In order to avoid packet duplication from the "basic suppression scheme" presented above other types of suppression have been proposed as identified in [7]:

(1) *Area-Based Suppression* (prevents only the packet duplications caused by nodes not being in transmission range of each other)
(2) *Active Selection* (prevents all forms of packet duplication, such as due to node movement, but at cost of bandwidth)

The idea in *area-based suppression* algorithm is to reduce the area from which the next hop is selected. This is done so that all nodes within this "suppression"

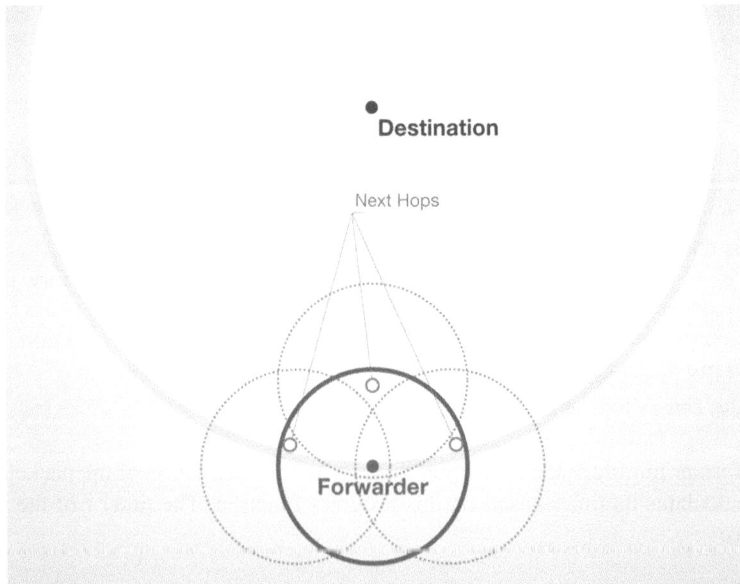


**Fig. 5.6** Forward the same message from multiple nodes [7]

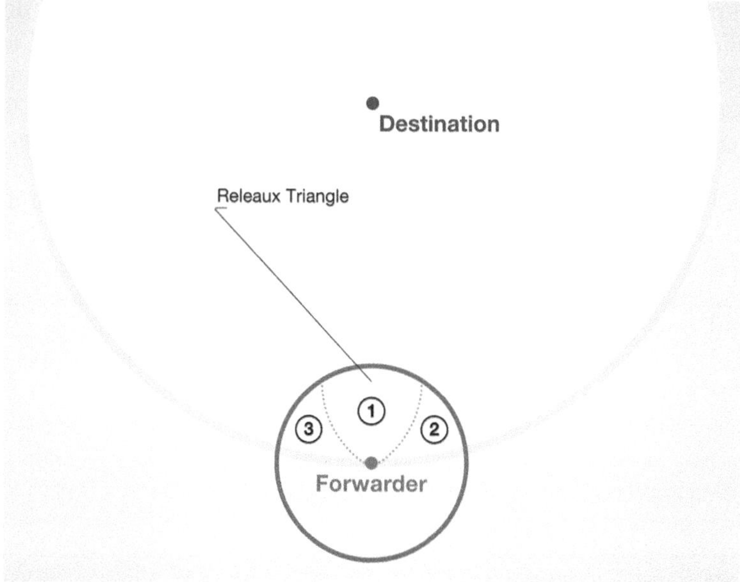


**Fig. 5.7**   Reuleaux triangle algorithm [7]

area are in transmission range of each other. The best decision on how the suppression area is chosen is the Reuleaux triangle (better than a circle!), where any two points are no further apart than the transmission range (see Fig. 5.7). To increase the chances of choosing a good forwarder (cover more neighbors), the Reuleaux triangle is used, rather than the circle.

The algorithm that uses Reuleaux triangle as suppression area has the following steps:

- Retransmit of the packet by the forwarder node;
- Participation in the contention process of the nodes within the Reuleaux triangle (1);
- Selection of the next forwarder as well as transmission to it, based on timer expiration;
- Suppression of the others nodes

If none of the neighbor nodes within areas (1), (2), or (3) responds to know a forwarder destination, a recovery strategy is required.

The main purpose of *area-based suppression* is to reduce the packet duplication, but this implies additional load on the network because a forwarding packet may be transmitted up to three times. However, the probability of requiring more than one transmission lowers as the number of nodes increases. Furthermore, the highest probability of finding a potential next forwarder is within the Reuleaux triangle because it covers the biggest of the three areas (1), (2) and (3).

The purpose of *active selection* is to avoid different forms of packet duplication using additional control messages, inspired by Request To Send, Clear To Send (RTS/CTS) scheme. The process is described next. The forwarder transmits a control packet, Request To Forward (RTF), which contains the location of the forwarding node as well as of the final destination node. The forwarding node also sets a reply timer according to the basic suppression scheme. When the timer expires, a control packet, Clear To Forward (CTF), which contains the location of the forwarding node is broadcasted. The suppression algorithm refers to the fact that when a node receives a CTF, it deletes its own timer. If the forwarder receives multiple CTF packets it selects the node with the largest forward progress and sends the information packet to this node via unicast. Furthermore, the advantage of active selection, as opposite to area based suppression, is that it may be integrated with RTS/CTS schemes to minimize the "hidden terminal" problem.

As opposed to the traditional position-based routing, where excessive re-sending of data may occur due to outdated neighbor table entries, the contention-based forwarding approach eliminates this disadvantage.

### 5.1.6 Octopus

Octopus separates the network space into horizontal and vertical strips [34]. Nodes in the same strip store each other's location. This location update technique is called synchronized aggregation and provides a good tradeoff between load (overhead) and fault-tolerance. The synchronization refers to the fact that no duplicate location updates are sent.

The octopus routing is designed for high mobile networks with nodes coming and leaving, which makes it suited for vehicular communication. It is a position based routing algorithm, meaning that each node determines its own position, e.g. using GPS. The algorithm provides recovery from faults such as lost packets due to MAC-level collisions or barriers.

Octopus is composed of three services: location update, location discovery and forwarding. The location update keeps two types of information for each node: radio range of the neighbors and neighbor nodes direction (north, south, west and east) in regard to the strip. The location discovery service discovers a target location when a node issues a location query. The last service, forwarding, forwards data packets to the target location.

The neighbors are updated upon receiving the beacon messages from every node (vehicle). This beacon messages contain ID, position and speed. A vehicle removes a neighbor if it does not hear from it for a set timeout such as 2 or 3 times the frequency rate of the beacons.

In theory, the location update protocol achieves 100% reliability. However, as identified in [34] under frequent failures and movements, the PDR lowers to 95%.

Location discovery is based on a query and a reply with information about the target towards the source. If permanent beacons are transmitted by all the nodes, no discovery service is required.

The geographic forwarding is based on the fact that each node knows one-hop neighbors and their locations. Each intermediate node forwards the packet to its neighbor that is geographically closest to the destination node. But in case the intermediate node is closer to the destination than all its neighbors the algorithm fails. This is why the routing algorithm chooses an alternative target, providing recovery from failure. Because of the moving nodes while packets are transmitted, the target's location has to be continuously re-estimated.

Based on different simulations [33, 34, 37] the channel load is low (due to geographic forwarding and single packet updates of the location of many nodes). It provides a good PDR, but it lowers in high density environment. The algorithm is position-based, fault-tolerant and uses a hybrid (proactive and reactive) approach.

### 5.1.7 Advanced Greedy Forwarding (AGF)

The Advanced Greedy Forwarding (AGF) improves the performance of GPSR for VANETs [14]. It is a technique that comes "over" a regular forwarding protocol such as GPSR. Note, that the AGF may be applied to any routing protocol based on flooding or greedy techniques.

The routing protocol was tested using simple information such as velocity vector information (speed & direction) and identification ("Hello") messages that are broadcasted by each node. Based on a reactive location service, the directions and speeds of two vehicles may be exchanged between them. Besides the identification messages, the header contains the time stamp. This time stamp is placed by each forwarding node.

The forwarding process works as follows: after receiving a data packet, a node must verify the neighbor table for the destination. If listed, it must be verified for validity. Another must is to take in consideration the packet travel time and the velocity vectors of the forwarding node as well as the destination node.

If the neighbor table contains the destination, but it is estimated as out of range, then the closest node to the destination is chosen as next hop. On the other side, if the destination is not in the neighbor table the node estimates if the destination can be reach within one hop based on packet travel time. If not even a potential destination exists within one hop, the non-propagating broadcast is sent around to the next closest node to the destination node, and the process repeats until the destination is finally found. A basic algorithm can be found in [14].

### 5.1.8 Preferred Group Broadcasting (PGB)

The Preferred Group Broadcasting (PGB) is a routing technique [14] that comes on top of an existing regular forwarding protocol, such as AODV, that reduces the overhead of control messages by removal of redundant broadcasts. It also enhances the routes providing autocorrect function. This results in higher stability in high mobile ad hoc networks, which is a key element for vehicular communication.

The PGB algorithm eliminates undesirable situations such as:

- too short distances between hops (too many hops in the path);
- too long distance between hops (poor connection quality, connection may be lost);
- the "hidden terminal" problem, detailed in Chap. 6 (This refers to the adjustment of the NAV (Network Allocation Vector) of the nodes S – source and R – receiver not to interfere, while other nodes may interfere by starting a transmission).

The routing load is reduced by the restriction in the PGB algorithm so that only certain nodes rebroadcast. This refers to separation in three groups for each node based on power signal level (which is generic presented as circles in Fig. 5.8, although in reality the algorithm doesn't consider the range circular):

- IN group represents the area of nodes with a stronger signal than in PG;
- PG (Preferred group) represents the area of the preferred set of nodes;
- OUT group represents the area of nodes with a weaker signal than in PG.

The received power signal is compared with two values, the Inner Threshold (IT) and Outer Threshold (OT). This way the node can be classified. The Receiving Threshold, noted of the figure as rxThresh (shortly rxTh), represents the power of a signal that corresponds to the maximum transmission range.

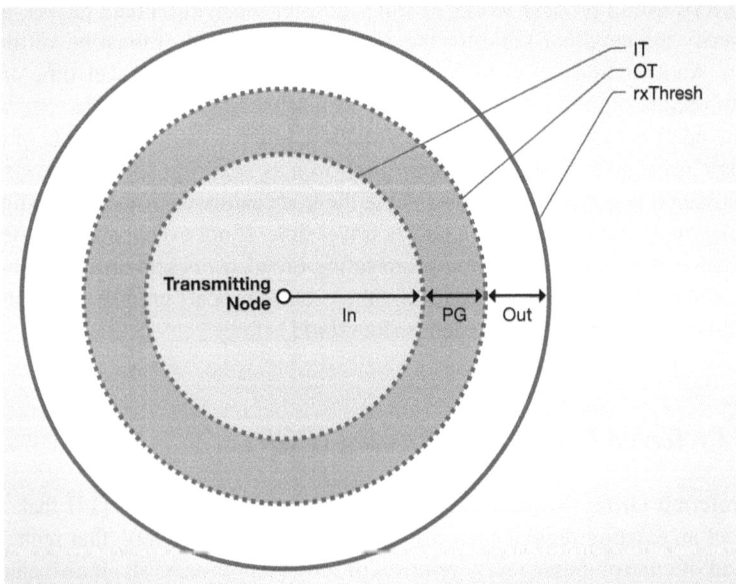


**Fig. 5.8**  PGB algorithm [14]

The relation between IT, OT and rxTh are:

$$IT = rxTh + fIT$$

(12) Power of received signal

$$OT = rxTh + fOT$$

(13) Power of received signal
where fIT , fOT are positive integer values, measured in dB. They are adjusted according to the neighborhood environment (traffic).

Depending on the sensed signal power a node determines if it is part of the IN or OUT group. It is IN, if the power is greater than IT, OUT otherwise. Each group carries an ID accordingly. The PG has ID=1, then the OUT group that has the ID=2 and finally the IN group with ID=3. The lowest ID (1) has the highest priority to be selected as forwarder.

Furthermore, the priority of a node may also be set based on the coordinates (besides sensed signal power).

The power range depends on the fIT and fOT and namely is defined as:

$$\Delta P = IT - OT = fIT - fOT$$

(14) Power range
By changing the value of fIT directly influences the number of neighbor nodes to be kept in the close-up power range of each node. By changing fOT we can modify the information dissemination traffic (this is generally set once at the beginning).

To account for differences in transmission and reception properties of the mobile nodes, the PGB algorithm defines two additional fields need to be included by each sending node: the transmission power (txP) and the receiving threshold (rxTh). Based on these fields, the receiving nodes are able to estimate the maximum transmission range, as well as to detect the signal loss.

Before retransmitting a packet a node waits until its hold off period expires. The hold off period depends on the group the node belongs to, namely: the shortest hold off time has the group with the highest priority. Using a GPS or another positioning system, the decision-taking process can be optimized. This refers to the fact that each node listening to other broadcasts during the hold off time is able reconstruct positions of each sending node and each sending node's predecessor.

To avoid lost RREQs, each node that retransmits a RREQ adds to the packet an ID and location of the predecessor as well as current node's location. The node, knowing it local position, is able to determine the triangle height for each received copy of the same RREQ. If the triangle height is lower than a certain threshold (which is a customizable parameter) the packet is dropped.

Normally, the nodes from the OUT group only verify that the transmission from PG was successful and further forwarded. But if the transmission from the PG does not reach any further nodes (including bad connection) then a node from the OUT group will retransmit the packet.

The packet is eventually retransmitted from the IN group nodes if there are no nodes in the PG and OUT groups. If a node from IN group receives the same transmission from another node with a different predecessor or the triangle height is lower than a certain threshold, it will drop previously received packet.

The authors of [14] propose an automatic route optimization by allowing splitting and merging of hops in the PGB algorithm. This optimization is beneficial in keeping the broadcast from being interrupted and avoiding redundant retransmissions.

Hop splitting is done when the distance between two nodes increases, so that the connection quality becomes poor. This is done by every intermediate node that overhears these transmissions and compares the power levels of the received signals from both nodes. It checks if these powers are high enough so that the current node can effectively communicate with both nodes. If so, the new path is generated.

Hop merging is done when the distance between nodes gets too short and one or more hops can be skipped without the loss of signal quality. The nodes need to update their routing tables.


## 5.2 Secure Multi-hop Routing

The following sections present modifications and extensions to existing multi-hop routing protocols to provide secure routing, i.e., the protection of the routing mechanisms and the routed data against manipulation, and eavesdropping.


### 5.2.1 Authenticated Routing for Ad Hoc Networks (ARAN)

ARAN resembles to AODV [8], but provides additional advantages such as message integrity and non-repudiation using pre-determined cryptographic certificates. The ARAN algorithm is designed to protect the network from malicious attacks in an ad hoc network (with no infrastructure to rely on). The basic principle of ARAN is the use of a trusted certificate server with a public key. Each node retrieves a certificate from the server to authenticate itself to other nodes. This certificate contains the IP address of current node, its public key, and a timestamp of creation and expiration of the certificate.

A route discovery packet (RDP) may be trigged by each node. In the packet are encapsulated the following information: destination node, its certificate, a nonce N, and a timestamp t, all digitally signed with current node private key. The so-called nonce is used to uniquely identify an RDP. Each node that forwards the packet verifies the previous node's signature and then removes it along with its certificate. Each forwarding node, also records the IP address of the previous node; afterwards it sign the packet and applies its own certificate.

Upon reception of a RDP, the destination unicasts a signed reply packet (REP) back to the destination on the same path, similar to AODV algorithm. Then the same technique is repeated (each node validates the previous node's signature, and so on).

One route is deleted from the routing table when no transmission occurs in the particular route's lifetime, similar to AODV algorithm. A recovery from failure is implemented in the form of a signed error message (ERR) that travels back to source to announce a disrupted route.

To summarize, ARAN prevents spoofing of the route, alteration of messages (e.g.: alteration of TTL values), and fake route requests (via nonce and timestamp). From the side of AODV performance the authors establish that ARAN, even though add security features, shows a good performance in discovering and maintaining routes. Unfortunately, the overall performance is poor. Main reasons are the inherent and – with the approach of ARAN aggravated – problems in AODV, and the increase packed and processing overhead due to digital signatures for every packet.

## 5.2.2 *Secure Ad Hoc on Demand Vector (SAODV)*

SAODV is a secure extension for AODV [8]. The main objectives of SAODV, like ARAN, are integrity, authentication, and non-repudiation of AODV routing information. The security mechanisms used are digital signatures and hash chains. The digital signatures are used for authentication of non-changeable part of the messages, while hash chains secure the changeable part of messages (the number of hops).

The hash function uses the random number generated when a RREQ or a RREP message is sent as well as the identifier. Besides them, a maximum hop count is also set (according to AODV algorithm) which is also included in the hash function. In this way, the receiving nodes are able to check the number of hops for each message.

Digital signatures are also required to guarantee the integrity of non-changeable part of the messages, by signing everything, except for the hop count and the hash field.

Although both security mechanism presented above are good in security, if the route is too new, the intermediate nodes cannot reply to RREP messages (according to AODV algorithm). To overcome this, two possible solutions exist. One is to ignore the route, so that each node just forwards the RREQ independent of the route. The second solution is to keep the route, but when a node generates a RREQ it will also include the RREP flag, prefix size and signature of the message so that intermediate nodes are able to route on the specific path to destination. However, the second solution requires additional overhead.

SAODV is an extension of AODV, resulting in similar performance characteristics; like in the ARAN approach, the effects of known problems in AODV increase, in particular since the processing of each packet requires more time, due to the usage of asymmetric cryptography.

### 5.2.3  Secure Link State Routing Protocol (SLSP)

SLSP provides secure proactive routing [8]. It uses one-way hash chains (to keep the hop count information) and asymmetric cryptography to protect topology discovery and the distribution of link state information. SLSP requires the use of a trusted certificate that may be provided by the "coalition" of some nodes or a third-party server.

In order to protect the topology discovery and the distribution of Link State Updates (LSUs), SLSP takes a series of actions in the ad hoc network:

- eliminate or restrict inexistent, fabricated links;
- prevents nodes of masquerade their peers;
- consolidates the robustness of neighbor discovery;
- protects against flood attacks, which is done by "rating nodes" mechanism. This is based on prioritization of nodes (higher priority for less LSUs generated).

The nodes keep topological information about their neighbors up to a defined number of hops. These linked nodes form a zone. The nodes also set and sent their LSU within a zone. A one-way hash function that authenticates the hop count is used to prevent packets from being propagated outside its zone. The nodes have to transmit regularly their public keys within the zone in order to keep up to date keys of the new ones when topology changes.

SLSP defines a Neighbor Lookup Protocol (NLP) with the following three main functions:

- keeping a MAC and IP addresses mapping of the neighbor nodes;
- identifying inconsistencies (e.g.: multiple IPs for a single data link interface)
- evaluating the traffic of control packets received by each neighbor (protection against DoS attacks).

Since the protocol makes use of asymmetric cryptography, it also requires some additional processing power.

### 5.2.4  Secure Position Aided Ad Hoc Routing (SPAAR)

SPAAR uses location information in order to provide better efficiency and security in mobile ad hoc networks. Designed with security as a top requirement, it uses geographical information to make forwarding decisions. Similar to CBF technique, the SPAAR reduces the number of routing messages. For security, the algorithm makes use of the asymmetric cryptography to protect and if not possible, to minimize the attacks from compromised nodes.

Each node in the network requires:

- a global pair of keys, one public and one private;
- a local pair of keys, one public and one private to communicate with its neighbors;

- a trusted certificate server;
- a public key of the certificate server.

The algorithm provides integrity, authentication, and non-repudiation.

Each node is able to send RREQ (route request) or a RREP (route reply) message like in AODV algorithm. Each of these messages are signed with the local private key and encrypted with the local public key of a neighbor. Every node can verify that the message was sent by a one-hop neighbor. On the other hand, the destination can also verify the identity of the sender.

This method requires that each node sends/stores a series of content information:

- a table for one-hop neighbors;
- a table for the last destinations it has transmitted packets to;
- periodic identification ("hello") messages to quickly detect topology changes;
- location request;
- location reply.

The only difference between the above two tables is that the destination table has to save information about the speed of the node to predict its next position as opposite to the first table, where the position is directly exchanged through table update messages.

A node broadcasts a location request when the node doesn't have an entry for destination. If any node that receives the message knows the location position of the required node, it replies with a location reply.

The mechanism is very efficient in security, but when applied, it requires the double of processing time, since it uses asymmetric cryptography for end-to-end communication as well as for hop-to-hop communication. However, the overall overhead is reduced due to the use of the geographic routing.

# References


1. T. Kosch, Technical concept and prerequisites of car-to-car communication, München: BMW group research and technology, http://www.car-to-car.org
2. M. Torrent-Moreno, F. Schmidt-Eisenlohr, H. Füßler, and H. Hartenstein, Packet forwarding in VANETs, the complete set of results, Karlsruhe, Germany, 2006
3. X. Ma and X. Chen, Performance analysis and enhancement of safety applications in DSRC vehicular ad hoc networks, 2007
4. A. Quintero, D.Y. Li, and H. Castro, A location routing protocol based on smart antennas for ad hoc networks, 2006
5. D. Gada, R. Gogri, P. Rathod, Z. Dedhia, N. Mody, S. Sanyal, and A. Abraham, A distributed security scheme for ad hoc networks
6. P.M. Melliar-Smith and L.E. Moser, Progress in real-time fault tolerance, Santa Barbara, CA, IEEE 2004
7. H. Füßler, J. Widmer, M. Mauve, and H. Hartenstein, A novel forwarding paradigm for position-based routing (with implicit addressing), Germany, 2003

8.  E. Fonseca and A. Festag, A survey of existing approaches for secure ad hoc routing and their applicability to VANETS, NEC network laboratories, March 2006
9.  C. Lin, AODV routing implementation for scalable wireless ad-hoc network simulation (SWANS)
10. C.E. Perkins et al., Ad hoc on demand distance vector (AODV) Routing, RFC 3561, July 2003
11. X. Jiang and T. Camp, A review of geocasting protocols for a mobile ad hoc network, Colorado, 2002
12. K. Kuladinithi, A. Timm-Giel, and C. Görg, Mobile ad-hoc communications in AEC industry, 2004
13. Y. Sakurai and J. Katto, AODV multipath extension using source route lists with optimized route establishment, Japan, IEEE 2004
14. V. Naumov, R. Baumann, and T. Gross, An evaluation of inter-vehicle ad hoc networks based on realistic vehicular traces, Zurich, 2006
15. C. Lochert, H. Füßler, H. Hartenstein, D. Hermann, J. Tian, and M. Mauve, A Routing strategy for vehicular ad hoc networks in city environments, Germany, 2003
16. V. Naumov, R. Baumann, and T. Gross, An evaluation of inter-vehicle ad hoc networks based on realistic vehicular traces, ACM 2006
17. K. Sanzgiri, D. LaFlamme, B. Dahill, B.N. Levine, C. Shields, and E.M. Belding-Royer, authenticated routing for ad hoc networks, IEEE 2005
18. A. Ho, Y. Ho, and K. Hua, A connectionless approach to mobile ad hoc networks in street environments, USA, 2004
19. Y. Ho, A.H. Ho, K. Hua, and G. Hamza-Lup, A connectionless approach to mobile ad hoc networks, Florida, USA. IEEE computer society, Los Alamitos, CA, 2004
20. Y. Ho, A.H. Ho, K. Hua, and T. Do, Adapting connectionless approach to mobile ad hoc networks in obstacle environment, Florida, USA, IEEE 2006
21. B. Zhou, Y.Z. Lee, M. Gerla, and F. de Rango, Geo-LANMAR: a scalable routing protocol for ad hoc networks with group motion, Wiley, New York, 2006
22. G. Mulas, A. La Piana, F. Cau, P. Iodice, A. Tamoni, and S. Antonini, A multi-hop MANET demonstrator tested on real-time applications, IEEE, 2005
23. S.M. Das, H. Pucha, and Y.C. Hu, Performance comparison of scalable location services for geographic ad hoc routing, IEEE, 2005
24. V.C. Giruka and M. Singhal, Hello protocols for ad-hoc networks: overhead and accuracy tradeoffs, IEEE, 2005
25. B. Karp and H.T. Kung, GPSR: greedy perimeter stateless routing for wireless networks, MobiCom, 2000
26. J. Li, J. Jannotti, D.S.J. De Couto, D.R. Karger, and R. Morris, A scalable location service for geographic ad hoc routing. In Proceedings of the MOBICOM'2000, Boston, 2000
27. M. Käsemann, H. Hartenstein, H. Füßler, and M. Mauve, Analysis of a location service for position-based routing in mobile ad hoc networks, Germany, 2002
28. H. Hartenstein, M. Käsemann, H. Hartenstein, H. Füßler, and M. Mauve, A simulation study of a location service for position-based routing in mobile ad hoc networks, 2002
29. B.A. Ihle, Grid location service, BTU Cottbus, 2004/2005
30. A. Helmy, G. Grewal, S. Pradhan, S. Sharma, and V. Alatzeth, A comparative study of mobility prediction schemes for grid location service (GLS), California
31. N.K. Guba and T. Camp, GLS: a location service for an ad hoc network
32. H. Hartenstein, M. Käsemann, H. Hartenstein, and H. Füßler, A reactive location service for mobile ad hoc networks, 2002
33. I. Keidar, R. Melamed, and Y. Barel, Octopus: a fault-tolerant and efficient ad-hoc routing protocol, Technion, 2005
34. R. Melamed, Scalable services for dynamic wide-area environments, 2006
35. M. Käsemann, H. Füßler, H. Hartenstein, and M. Mauve, A reactive location service for mobile ad hoc networks, Germany, 2002
36. C.J. Adler, Information dissemination in vehicular ad hoc networks, München, 2006
37. L. Itkin, E. Gurevich, and I. Vaisband, OCTOPUS scalable routing protocol for wireless ad hoc networks, 2005

# Chapter 6
# Medium Access for Vehicular Communications

MAC (Medium Access Control) layer protocol consists of a set of rules, so that a node knows when to transmit messages and when not to. The message has a lifetime and during this lifetime the packet is transmitted and retransmitted. Afterwards, the message is discarded [1, 2]. The MAC protocol is used to combat the collision problem at the receiver. The design of the protocol has the goal to achieve high reception reliability and low latency [1].

A key factor is the time stamp of each packet transmitted and received. A reference signal for time synchronization is required. Two different time synchronization approaches may be used [3]:

- GPS-based. This one uses a globally known time information coming from the satellites. The rubidium or cesium oscillators on board of the satellites are referenced to Coordinated Universal Time (UTC) and provide a highly reliable time reference source with an absolute error significantly below 1 $\mu$s. Of course, this accuracy is provided only as long as GPS signal is not disturbed, otherwise accuracy is provided by the stability of the local oscillator.
- Decentralized. This one uses a mutual adaptation of the individual node timing. According to [3] a mechanism is proposed that can avoid a "normal" systematic timing drift (for a distance <1,000 m an error below 5 $\mu$s is observed).

A specific MAC layer is used for bidirectional and other for position based communication regime. They are further detailed below.

In the classic bidirectional MAC (Fig. 6.1), initially, the initiator (vehicle X) selects its responder (vehicle Y) based on necessity. In the control channel, the handshake starts from the sender by sending a short Request-to-Send (RTS) packet to the intended receiver. Then, the receiver replies with a Clear-to-Send (CTS), while the sender begin to transmit data after it receives the CTS packet. The goal of RTS-CTS handshake is to notify the neighbors of both the sender and the receiver as well as to prevent sending data to the receiver before it is ready. Although this is called the CSMA/CA (Carrier Sense Multiple Access with Collision Avoidance) scheme [4], it only reduces collisions, but does not eliminate them. After the handshaking, the initiator sends the actual information and waits for acknowledge from responder for confirmation. The responder may also transmit data to the initiator,

R. Popescu-Zeletin et al., *Vehicular-2-X Communication*,
DOI 10.1007/978-3-540-77143-2_6, © Springer-Verlag Berlin Heidelberg 2010

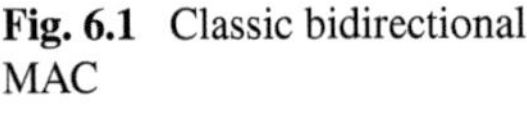

**Fig. 6.1** Classic bidirectional MAC

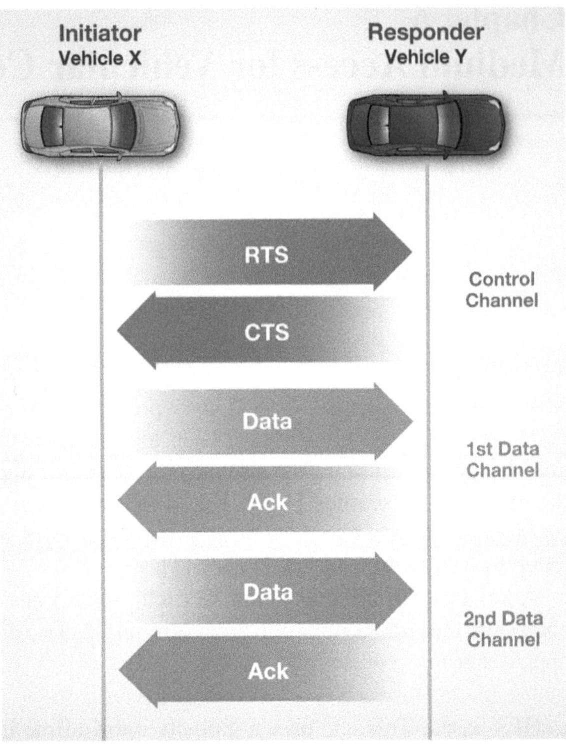


as the transfer of information is bidirectional. Two data channels are used to enable simultaneous bidirectional transfer [5]. This kind of transfer is called full-duplex or frequency-division duplex.

The classic bidirectional MAC is used for non-critical features such as provider and adhoc services.

The fast bidirectional MAC (Fig. 6.2) eliminates RTS/CTS in order to provide a faster connection in critical safety applications. Therefore the control channel contains only of a Fast Bidirectional Transmit (FBT) packet that announces the responder to open a fast bidirectional connection in case of necessity (e.g. imminent accident). Same as above, the two data channels are used to enable fast, simultaneous bidirectional transfer (full-duplex).

In the position based regime the receiver feedback would create new communication problems such as delay and overhead, therefore acknowledgment of the information is skipped (Fig. 6.3) [2]. When there are many receivers or the network is highly mobile, learning identity itself requires significant bandwidth. To provide communication as quickly as possible, the vehicles do not provide authentication and association procedures.

The messages have a specific update rate, which implies that the reception has a particular probability [1]. This is the case of position based regime.

Two types of messages are being sent this way: alert type and permanent beacon type.

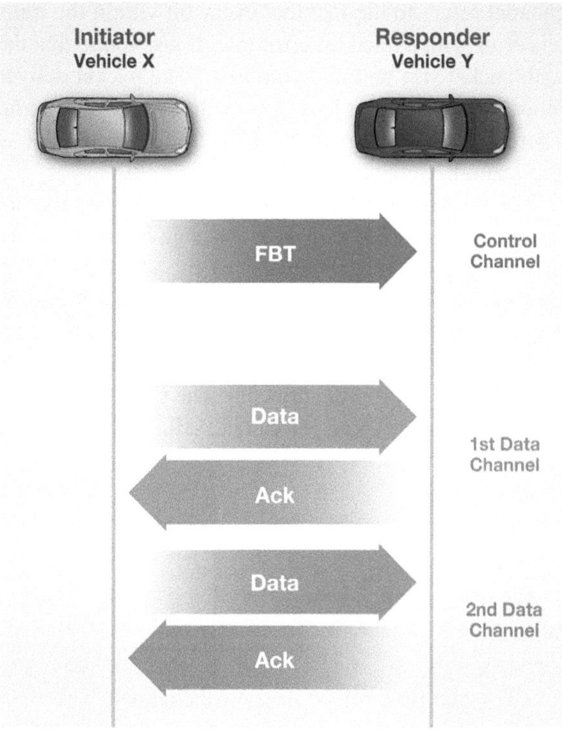

**Fig. 6.2** Fast bidirectional MAC



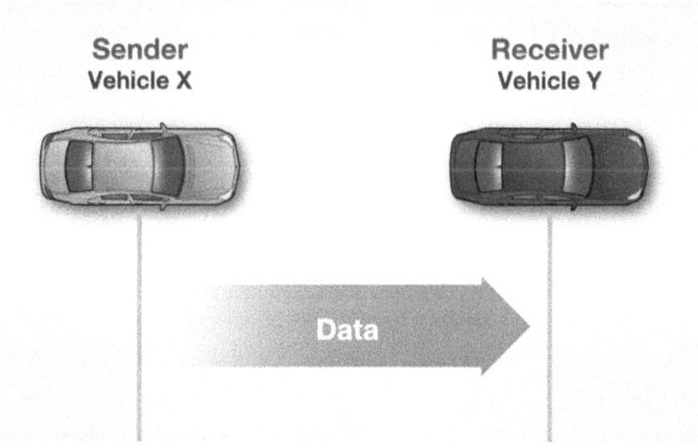

**Fig. 6.3** Position based MAC



In the vehicular network [6], since multiple mobile nodes share the same media, access to the common channel has to be made in a distributed way, taking in account mobility, hidden terminals and exposed nodes problems. All these problems are further presented below.

There are two performance measures, i.e., *reception probability (packet delivery ratio)* and *channel busy time*.

The reception probability is affected by different parameters: range, nodes mobility, nodes density, the imperfection of the channel [7]. The imperfection of the channel refers to the fact that every bit within the transmitted data packets encounters error with a fixed bit error rate. It was found that the fixed bit error rate increases with packet size and node mobility. The packet delivery ratio (PDR) must be theoretically 1 (no packet loss), but without acknowledgment this would be impossible. Therefore, according to [8, 9, 7], a minimum PDR of 0.95 is required for safety application domain.

The inverse of PDR (packet delivery ratio) is the probability of failure PF(L,t) in an ad hoc network [2] for a given range L, and message lifetime t, is the probability of a random message transmitted by a random vehicle that will not be received by a random receiver vehicle at distance L within time t. The interference (PF) is proportional with the distance between transmitter and receiver.

Channel busy time (CBT) is an important performance measure when operating in the DSRC control channel [1]. The CBT is defined as the time fraction in which the channel is occupied (by a successful or collided packet) and may not be used by other applications.

$$\mathrm{CBT} = \frac{\mathrm{T_{safety}}}{\mathrm{T}}$$

(15) Channel Busy Time
Where:

> $T$ – total time period in control channel;
> $T_{safety}$– total time periods within T occupied by safety messages.

Ideally, PF and CBT should be low [2].

Vehicle safety applications generate a message to be transmitted to other vehicles [1]. The message is passed down to the MAC layer. The MAC protocol, a set of rules by which a vehicle decides when to transmit its messages and when to keep silent, attempts to broadcast the packet only within the message's lifetime and discards the packet when the lifetime expires [1, 2]. The MAC protocol is used to combat the collision at the receiver problem.

The Fig. 6.4 shows the concept of repetitive transmission [2]: at each transmitter the protocol evenly divides the message lifetime into n slots. The number of slots (n) is:

$$n = \left\lfloor \frac{t}{t_{trans}} \right\rfloor$$

(16) Number of slots
Where:

> $\lfloor \ \rfloor$ – is the maximum integer of the argument;
> $t$ – lifetime;
> $t_{trans}$ – time needed to send one packet.

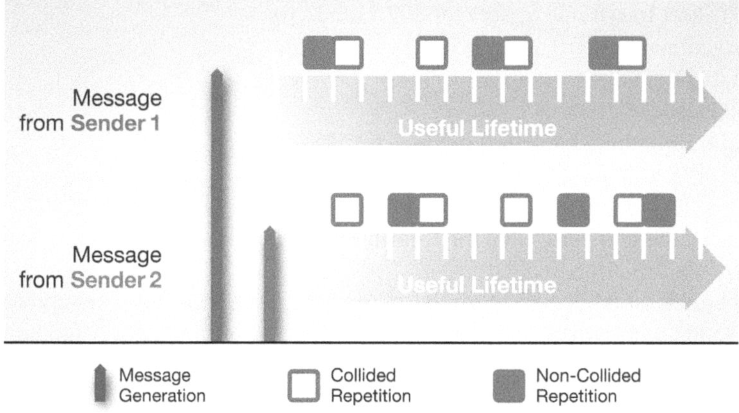



**Fig. 6.4** Collision at the receiver problem [2, 10]

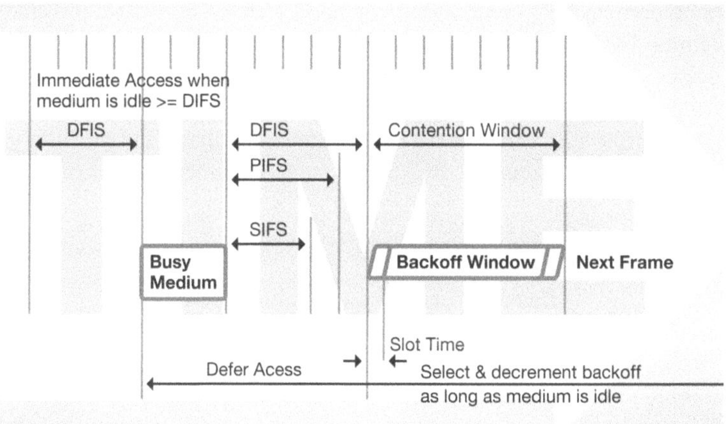



**Fig. 6.5** DSRC MAC scheme [7, 13]

If at least one of the packets is received without collision implies that the message is successful received by the given receiver. The message transmission fails if all of its repeated packets are lost due to collision. In order to prevent a collision [7, 11, 12], the 802.11a MAC layer, which is the base of DSRC MAC layer, uses the distributed coordination function (DCF). The DCF is a random access scheme for all associated devices in a cluster (termed the *basic service set*) based on carrier sense multiple access with collision avoidance (CSMA/CA). The broadcast of DSRC MAC is presented in Fig. 6.5.

The parameters of DSRC MAC layer as identified in [7]:

– Slot time, $\sigma$: 16 $\mu$s
– DIFS (Distributed Interframe Space) for 802.11a: 64 $\mu$s (SIFS+2* $\sigma$)

- SIFS (Short Interframe Space) for 802.11a: 32 $\mu$s
- Propagation delay, $\delta$: 1 $\mu$s
- Preamble Length: 40 $\mu$s
- PLCP Header Length: 8 $\mu$s
- CWMin (minimum backoff window size): 15
- CWMax (maximum backoff window size): 1,023

Some of the parameters are different from the 802.11a MAC layer as identified in [13].

The broadcast procedure in DSRC follows the basic medium access protocol of DCF [7, 13] with no positive acknowledgement and retransmission functions. As a result, current backoff window size is always a constant. In DCF scheme, before transmission each node checks first if the medium is busy or not with the help of a carrier sense mechanism. The transmission is started immediately, if the medium is idle. On the other hand, when the medium is busy, the node has to wait until the medium is idle without interrupting the DIFS period. Then, as long as the medium is idle, a random backoff process is conducted by choosing a random value for the backoff timer (BC). When the backoff process is over, the transmission is activated. BC is decremented by 1 for every slot time the medium remains idle. When the backoff timer reaches the zero value, the transmission starts. During the backoff countdown process the carrier-sense persists, meaning that if the medium is sensed to be busy the BC is frozen. For BC to be resumed the medium has to be idle for a DIFS period.

Thus the backoff time [13] can be expressed as:

$$\text{Backoff Time} = \sigma \times \text{Random}[0, \text{CW}]$$

(17) Backoff time

The timing of the backoff process is kept by each node in a Contention Window (CW). Random[0,CW] is a random value chosen from the uniform distribution in the interval [0, CW]. The CW takes values from CWMin (initially) to CWMax. If the medium is idle for a SIFS period (shorter than DIFS) the receiver is able to strat receiving. After transmission of a frame the node will start another backoff procedure.

The CSMA/CA scheme used by DSRC MAC layer minimises the probability of a collision as long as the involved stations are in sensing range [7, 11, 12]. Unfortunately, this cannot be assumed for an ad-hoc network in which the hidden terminal problem can arise. The hidden terminal problem, more evident in multicast than in unicast, is illustrated in Fig. 6.6 [7, 11]. The issue is that nodes in the receiving region of vehicle(s) R, but not in the region of sending vehicle $S$ (shaded area), may cause the hidden terminal problem. Please note that Fig. 6.6 shows four receiver nodes to indicate a multicast environment.

In this situation vehicle S is broadcasting data. Vehicle(s) R is in reception range but not vehicle P. Since the latter is not aware of the ongoing communication, it

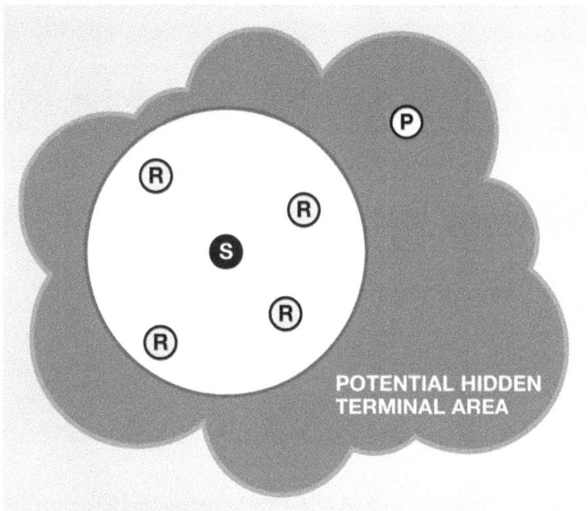

**Fig. 6.6** Hidden terminal problem in multicast environments [7, 12, 4]



may start transmitting data as well. This could result in a message collision between the two sending stations, namely at vehicle R. To the best of our knowledge, no efficient scheme was found to eliminate the hidden terminal problem in a multicast communication. Therefore collisions are assumed to occur.

Also, DCF is not capable of providing predictable quality of service (QoS) [7, 13]. The development of a robust and efficient MAC protocol is a requirement for the new generation of DSRC devices. A solution may be, as identified in [13] the use of an Enhanced DCF (EDCF). In EDCF, traffic with different QoS requirements is divided into different Access Categories (AC). The contention method of EDCF is the same as that of DCF except that each AC uses a different set of CW and DIFS values. In vehicular communication, unlike the traditional WLAN environment, some applications (such as the cooperative forward collision warning) impose a strict requirement on delay. Comparing DCF and EDCF in terms of transmission delay and throughput was found that EDCF gives better results. Unfortunately, no tests were conducting regarding the most important QoS: PDR (packet delivery rate).

The information from the messages is transmitted to other vehicles via packets [1]. The packet has to contain the sender position, the receiver vehicle's position, the type of the event, etc. The time taken to transmit one packet directly depends on the packet size and the transfer data rate.

The usage of message content can be very beneficial as identified in [10]. A measure that sets the priority of a message, called transmission relevance, computes a local order of the messages within the send queue of each node. Therefore, the basic mechanism is to give priority to a particular message in a shared medium. The contention is based on the relevance of messages. The most relevant message wins and it is sent first. An example of three vehicles that use relevance functions for their

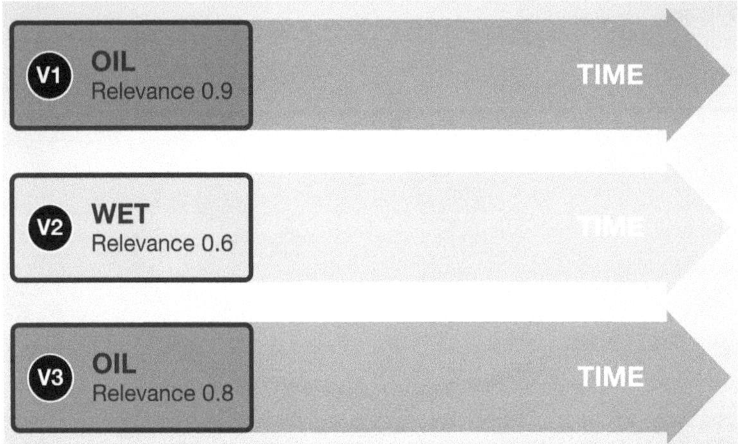


**Fig. 6.7**  Content-based on relevance of messages [10]

messages is presented in Fig. 6.7. In the example, vehicle V1 calculates the highest
relevance of 0.9 for the oil message. So the V1 vehicle has the greatest priority and
shortest backoff timer (the backoff timer is inverse proportional with the relevance).

The relevance-based approach brings the possibility to create a situation-adaptive
and self-organizing system that continuously adapts to the current context, which is
a great advantage in vehicle ad hoc networks. As a result, the overall system performance is better due to the control of the available bandwidth through the relevance
of the messages.

If a common channel, for alert type messages is shared by all vehicles and is allo-
cated on-demand, then a contention mechanism based on relevance may be required.
The mechanism has to decide which node has the right to access the channel at any
moment. For beacon messages, it is not possible to define priority, as these messages
are not indicating any events and all have similar relevance. And because predicted
transmission range is about 1,000 m, collision of messages on MAC layer is the
main challenge.

A solution to this problem is to use subcarriers that are transmitted simultaneously at different frequencies to the receiver as detailed in the next section.

# References


1. Q. Xu, T. Mak, J. Ko, and R. Sengupta, MAC protocol design for vehicle safety communications in dedicated short range communications spectrum, California USA
2. Q. Xu, T. Mak, and R. Sengupta, Vehicle-to-vehicle safety messaging in DSRC, Pennsylvania,
   2004
3. A. Ebner, L. Wischhof, H. Rohling, R. Halfmann, and M. Lott, Time synchronization in highly
   dynamic ad hoc networks

4. W. Ye and J. Heidemann, Medium access control in wireless sensor networks, October 2003
5. T. Kuang and C. Williamson, A bidirectional multi-channel MAC protocol for improving TCP performance on multihop wireless ad hoc networks, October 2004
6. J. Lee, Y. Tscha, and K. Lee, Medium access control protocol using state changeable directional antennas in ad-hoc networks, 2003
7. X. Ma and X. Chen, Performance analysis and enhancement of safety applications in DSRC vehicular ad hoc networks, 2007
8. M. Torrent-Moreno, F. Schmidt-Eisenlohr, H. Füßler, and H. Hartenstein, Packet forwarding in VANETs, the complete set of results, Karlsruhe, 2006
9. S. Yousefi, A. Benslimane, and M. Fathy, Performance of beacon safety message dissemination in vehicular ad hoc network
10. C.J. Adler, Information dissemination in vehicular ad hoc networks, München, 2006
11. A. Meier, 5.9 GHz dedicated short range communication – design of the vehicular safety communication architecture, 2005
12. J. Zhu and S. Roy, MAC for dedicated short range communications in intelligent transport system, IEEE, 2003
13. X. Xia and Z. Niu, Performance of EDCF Mac scheme for future multi-service DSRC based road-to-vehicle communication systems, in ITS, IEEE, 2004

# Chapter 7
# Physical Layer Technologies

Although IEEE 802.11b has been demonstrating some capabilities for the communication between mobiles at high speed in ITS (Intelligent Transportation Systems), a new standard was introduced: IEEE 802.11p [1]. The lower layer of IEEE 802.11p is the base standard for the new coming DSRC (Dedicated Short Range Communications), which involves vehicle-to-x communication. The frequency allocation in US (5,850–5,925 GHz) was done from 2004, while in Europe, EU DSRC was adopted in August 2008 with the frequency band within the range of 5,875–5,905 GHz [2]. Currently, a newly formed Wireless Access in Vehicular Environments (WAVE) study group works on the migration of IEEE 802.11 standards toward 802.11p [3, 4]. The WAVE study group is working on more standards: IEEE P1609.3 that specifies the overall communication architecture and the IEEE 802.11p, IEEE P1609.1, IEEE P1609.4, IEEE P1609.2 which focus on the architecture's details. The 802.11p PHY layer follows the same frame structure, modulation scheme and training sequences of the IEEE 802.11a PHY layer [3–5].

In comparison to cellular communications, DSRC can provide higher transfer rates and smaller communication latencies for small communication zones defined by the communication radius of the technology [6]. It will support communication between nodes that travel with a speed of up to 200 km/h [7–8]. Besides the above features that the new 802.11p needs to provide, the "normal" wireless issues like: *path loss and fading* needs to be minimized [3–5]. The path loss refers to signal strength variation due to environment: it can attenuate faster or slower than it does in free space. The fading refers to the multi-path effect which also affects the signal strength.

The DSRC physical layer uses an orthogonal frequency division multiplex (OFDM) modulation scheme to multiplex data [3–5]. The technology works by splitting the radio signal into multiple smaller sub-signals. These sub-carriers typically overlap in frequency, but are designed not to interfere with each other: sub-carriers are orthogonal to each other and they are separated using a Fast Fourier Transform (FFT) algorithm. Main reasons for using OFDM are its high spectral efficiency [5], its good performance in multi-path fading environments and the simple transceiver design. Besides reducing multipath interference, it can actually increase the signal strength by processing the reflected packets to increase gain. This technique also improves non-line of sight delivery (where the sender and the

R. Popescu-Zeletin et al., *Vehicular-2-X Communication*,
DOI 10.1007/978-3-540-77143-2_7, © Springer-Verlag Berlin Heidelberg 2010

receiver do not see each other). OFDM divides the input data stream into a set of parallel bit streams and each bit stream is then mapped onto a set of overlapping orthogonal subcarriers for data modulation and demodulation. All of the orthogonal subcarriers are simultaneously transmitted.

Orthogonal Frequency-Division Multiple Access (OFDMA) provides shared access to multiply users by assigning subsets of subcarriers to individual users. Theoretically sub-channelization can be made in thousands of sub-channels.

The *high mobility* of vehicles [3] can be the cause for two negative effects: failure to receive messages or packet errors. The first is because during transmission of safety-related messages, some of the receivers may move out of transmission range in regard to the sender. The second negative impact refers to high packet error rates and consequently lower channel capacity because high mobility makes worse Doppler spread on orthogonal frequency division multiplex (OFDM).

US DSRC will have six service channel and one control channel as identified in [3–5], making a total of 7 channels (10 MHz each) for supporting safety and non-safety applications. Messages have different priorities: low (non-safety messages such as the announcement of a map update, etc) or high (safety messages). These messages are to be used in the Control channel, which is monitored by all vehicles. In Service channels are to be transferred the actual applications such as transfer of a map update that was announced in Control channel.

Note, that in the physical layer, the warning alerts are sent on a different channel than the permanent beacons. Similar, the Bidirectional Info is sent on a different channel than the Bidirectional Autonomous information.

In order to support home and Hot Spot access (comfort applications), besides 802.11p also the 802.11a/b/g need to be supported [9]. This means the device has to support both frequencies in parallel.

In Table 7.1 the specifications for the some of the 802.11 standards [7, 10, 11, 4] are presented:

The transmission range can go over the above values but the bandwidth will suffer (for instance the 802.11a may support up to 350 m but at 6 Mbit/s) [11].

**Table 7.1**   IEEE 802.11 standard specifications

| Wi-Fi standards | Modulation | Frequency [GHz] | Channel Bandw. [MHz] | No. of all/non-overlap channels | Bandwidth (max) [Mbit/s] | Transmission range (outdoor) [m] |
| --- | --- | --- | --- | --- | --- | --- |
| 802.11a | OFDM | 5,725–5,850 | 20 | 12/8 | 54 | 30 |
| 802.11b | DSSS | 2,400–2,485 | 22 | 14/3 | 11 | 250 |
| 802.11g | OFDM | 2,400–2,483 | 22 | 14/3 | 54 | 250 |
| 802.11p US | OFDM | 5,850–5,925 | 10 (20) | 7/7 | 54 | 1,000 |
| 802.11p EU | OFDM | 5,875–5,905 | 10 | 7 | 54 | 1,000 |

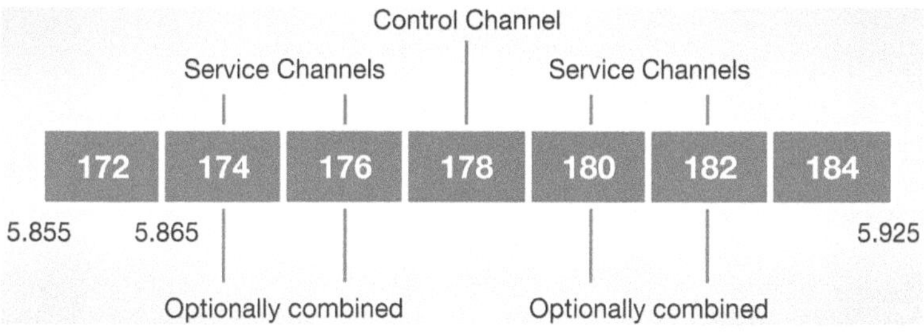


**Fig. 7.1** DSRC frequency band allocation in US [3, 12, 13]

The DSRC technical characteristics in the US are [3, 4, 12, 13, 5, 14, 15]:

- *Bandwidth*: 75 MHz (5,850 – 5,925 GHz)
- *Guard Channel:* 5 MHz reserved at the lower end
- *Channels*: 7 non-overlapping 10 MHz channels (Fig. 7.1). The channels are: *Control Channel*(178) is used principally to broadcast safety related communication only, while non-safety data exchange is strictly limited in terms of transmission time and interval. *Vehicle-to-vehicle channel* (172) and *vehicle-to-road channel or "Intersection" channel* (184) are used for dedicated safety-related applications (different from Control Channel). The remaining 4 service channels (174, 176, 180, 182), that may be combined in 2 channels of 20 MHz, are used for shared safety-related applications.
- *Modulation*: BPSK OFDM, QPSK OFDM, 16-QAM OFDM, 64-QAM OFDM. The first 128 bits are always BPSK coded. According to [3, 5] QPSK outperforms both BPSK and 16-QAM OFDM for all packet sizes.
- *Coding Rates*: 1/2, 2/3, 3/4
- *Subcarriers*: 64 (of which only 52 subcarriers are actually used for signal transmission)
- OFDM symbol duration: 8.0 $\mu$s
- *Data Rate*: 6, 9, 12, 18, 24, and 27 Mbps with 10 MHz Channels (3 Mbps preamble) (or 6, 9, 12, 18, 24, 36, 48, and 54 Mbps with 20 MHz Channel option) (6 Mbps preamble)
- *Power (Range)*: usually less than 33 dBm (2 W) but up to 44.8 dBm (30 W) for qualified public safety applications on the Control Channel which theoretically allow to broadcast safety messages up to 1,000 m. Only control channel (44.8 dBm) and "intersection" channel (40 dBm) are used for long range, up to 1,000 m ("special" messages from emergency vehicles or RSUs), the other channels are used for distances up to 330 m (collision avoidance, road condition). The power should be adjusted according to the speed of the vehicles to minimize interferences. Antennas are required. They may be mounted on the roof of the vehicle to obtain a 360° horizontal pattern and to be in the "center" of the vehicle to allow equidistant communication. Note that in Europe the standard of maximum total transmit power is limited to 33 dBm.
- *Minimum separation*: 15 m (on small zone channels)

The hardware components used in vehicular communication is composed OBUs (OnBoard Units), RSUs (RoadSide Units) and omni-directional Antennas, but also Sensors.

## References


1. G. Segarra, Activities and applications of the car 2 car communication: the Renault vision, http://www.car-to-car.org
2. http://www.car-to-car.org
3. X. Ma and X. Chen, Performance analysis and enhancement of safety applications in DSRC vehicular ad hoc networks, 2007
4. A. Meier, 5.9 GHz dedicated short range communication – design of the vehicular safety communication architecture, 2005
5. J. Yin, T. ElBatt, G. Yeung, B. Ryu, S. Habermas, H. Krishnan, and T. Talty, Performance evaluation of safety applications over DSRC vehicular ad hoc networks, USA, VANET, 2004
6. Q. Xu, R. Sengupta, and D. Jiang, Design and analysis of highway safety communication protocol in 5.9 GHz dedicated short range communication spectrum, USA
7. http://standards.ieee.org/board/nes/projects/802-11p.pdf
8. T. ElBatt, S. Goel, G. Holland, H. Krishnan, and J. Parikh, Cooperative collision warning using dedicated short range wireless communications, ACM, VANET, 2006
9. T. Kosch, Technical concept and prerequisites of car-to-car communication, München: BMW group research and technology, http://www.car-to-car.org
10. C. Ribeiro, Bringing wireless access to the automobile: a comparison of Wi-Fi, WiMAX, MBWA, and 3G, 2005
11. J. Moran, Wireless home networking – Wi-Fi standards, 2002, http://www.smallbusiness computing.com
12. J. Zhu and S. Roy, MAC for dedicated short range communications in intelligent transport system, IEEE, 2003
13. DSRC_Tutorial_06-10-021.ppt
14. J. Opiola and B.A. Hamilton, Vehicle infrastructure integration (VII) in the US – enhancing safety, enabling mobility
15. US Department of Transportation, Vehicle infrastructure integration (VII) – VII architecture and functional requirements version 1.1, 2005

# Chapter 8
# Security

We make the observation that all of the communication regimes require reliable security [1]. The main purpose of the comprehensive set of security mechanisms is to assure the physical safety of the passengers in the vehicle.

To define security, privacy and trust in the vehicular communication many aspects needs to be covered. Regarding security and trust, the system and its main assets must be protected against different types of malicious attacks.

The vehicle communication provides new application domains with great advantages to the drivers that were not possible before, but at the same time it opens the possibility for abuses and attacks. The negative impact on security of vehicle-to-vehicle communication system is the distributed ad hoc network created. This affects fundamental security building blocks, such as trust management and key distribution. One simple, but eloquent example to demonstrate that vehicular communication must be strongly secured is next presented. Imagine the consequences, if one vehicle is able to introduce false information in a large area of the vehicular network.

The constrains and objectives for security in vehicular network [2, 3] are scalability, mobility, low complexity and low cost. Scalability refers to the proper management of the network when the number of equipped vehicles increases (up to millions) with respect to security aspects. The main aspect to be resolved is the minimization of the overhead introduced by security system. Overhead introduced by security algorithms must also be small in order to provide fast communication in high mobility environment (relative velocities between vehicles of up to 400 km/h). The vehicles must determine the situation (exchange packets over a secured connection) in a rather small period of time. The low complexity factor refers strictly to the usage of the security system, which is desirable in the automotive domain. The simplicity also results in a lower cost, as opposite to a complex system that may induce higher costs.

The security mechanisms summarizes the following major aspects [1, 2, 4, 5]:

- security architecture,
- secure communication,
- privacy protection,
- trustworthy data and
- trustworthy system.

R. Popescu-Zeletin et al., *Vehicular-2-X Communication*,
DOI 10.1007/978-3-540-77143-2_8, © Springer-Verlag Berlin Heidelberg 2010

Security architecture identifies the major functions of a security system for a particular purpose. These purposes generally depend on the communication regime as well as application type.

Secure communication refers to cryptographic protection of the communication channel, including cryptographic primitives and key management.

Privacy protection of the participants means that the sending vehicles identity cannot be tracked back and the identity of the users is protected. The users should not be linked to the identifiers used for communication.

Trustworthy data refers to define a level of trust for the incoming information. This means that the receiving vehicles have a reliable mechanism to check that the messages have not been altered on their path and that the information in the messages is correct (message integrity). An example of crucial information is the mobility data in proactive applications.

Trustworthy system refers to define a level of trust for the environment. This means source authenticity (sender is a valid source) and detecting system manipulations. This means that the network must be protected against different types of attacks such as DoS (Denial of Service) or generating fake messages. DOS refers to make the resource unavailable to legitimate users. In particular, different authorization levels of the nodes are introduced (what is each node allowed to do in the network). For instance, a node may be limited to insert in the network only a specific type of messages.

As we already argued, low latency (represents time delay from the source vehicle to the destination) is required. Therefore, cryptographic methods that create less overhead are suited (e.g. asymmetric cryptography). A particular solution can be a PKI certificate that is issued for any node by a Trusted Third Party.

As identified in [6], asymmetric cryptography is the best available cryptographic scheme for vehicular communication network. Hovever, this adds advantages such as a reduction of key distribution complexity, and provision of non-repudiation. The author also concludes that the integration of a plausibility check technique would be beneficial when protecting against compromised nodes, since this technique wouldn't influence the overall performance of vehicular network, and incorrect information would be detected and corrected.

Local plausibility checks [2, 6] in vehicles include comparison of received information with the internal sensor data or evaluating messages from different sources about a single event and scenario. Simulations have shown that this substantially increases the effort of an attacker, but it requires proper models for each particular application. Beneficial for security is the ability of regular system checks on the nodes. This system check, also reduce the number of malfunctioning units and may update software (security fixes) if necessary.

A positive impact on security is also given by information about time and position [2]. To enhance security, restricted physical access to authorized personnel only in some key points, has to be introduced.

A security architecture [7] should define at least: relevant stakeholders, threats, vulnerabilities and risks, functional and non-functional requirements, system behavior and system components.

The relevant stakeholders are senders and receivers (vehicles, infrastructure) that use information such as audit data to verify that the messages are valid and not changed. Another group of stakeholders are the attackers.

Typically, they are identified and classified for the use in a threat, vulnerabilities and risks analysis (TVRA). Threats, vulnerabilities and risks are analyzed based on a defined system and its environment. This results in security objectives, and subsequently, functional requirements of the security system.

The non-functional requirements are overall system functionalities such as low cost or minimum latency requirement for a specific type of applications. The non-functional and functional requirements are interconnected to each other. For example, choosing the wrong security solution for proactive applications may introduce too much latency or overhead.

The components and their behavior of the system are two viewpoints of the security architecture. There are different security architectures for vehicular communication. The architectures are based on security services for ad-hoc communications. The general aspects are very common. Therefore we will illustrate a generic architecture in Fig. 8.1 such as the ones specified in NoW [2, 8–10], SeVeCom [11, 12], and GST [13] with GST-SEC subproject.

The five above security modules are defined for each node and forms the security layer. Each node (e.g.: vehicle) is connected with another through a secure tunnel (as presented in GST-SEC subproject).

The Communication Module, distinct for each communication regime, addresses secure communication in vehicular network.

The Trust Management Module manages trust and identity of the nodes involved. This module includes the use of a PKI for key and certificate management. The Cryptographic Storage & Processing Module provides a secure key/certificate storage, as well as high secure priority functions storage (e.g.: signature creation).

The Privacy Management Module protects the users of vehicles, offering them a certain level of privacy. This covers two aspects: first, to protect the link between

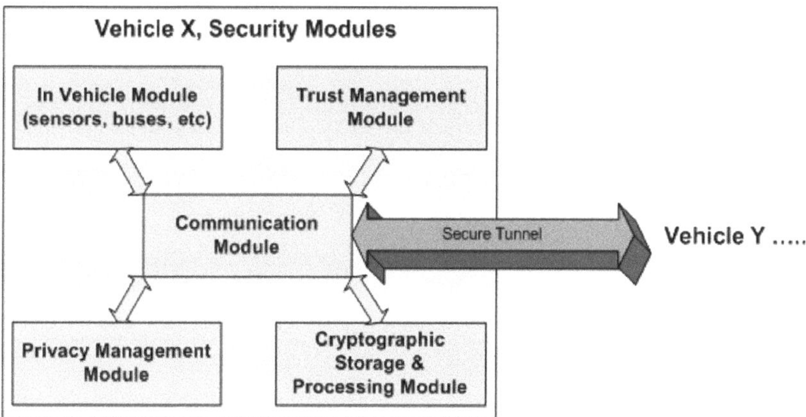


**Fig. 8.1** Conceptual security architecture [12]

the user and the vehicle and second, to prevent location tracking of the vehicle. Technically, this module leverages on pseudonyms (e.g.: certified public keys).

The In Vehicle Module assures protection against in-vehicle system (e.g.: sensors, buses, etc) manipulations. This module contains a firewall and intrusion detection. This also includes local plausibility checks with sensor information, position data, etc.

The noteworthy projects that cover secure vehicular communication are: SEVECOM, ERTICO GST (GST-SEC), Network on Wheels (NoW). These projects are further depicted in the Annexes. Although ended in 2008, the NoW [8, 10] was one of the first projects that addresses security and privacy in vehicular networks. Detailed work on position verification [2], location privacy and secure routing [6, 9]. The European project SEVECOM [11, 12], that also closed in the end of 2008, provides solutions for pseudonyms, key management and storage as well as secure communications. Other projects that started in 2008 are: EVITA (E-safety Vehicle InTrusion protected Applications) [14] and PRECIOSA (PRivacy Enabled Capability In cOoperative systems and Safety Applications) [15].

# References


1. T. Kosch, Technical concept and prerequisites of car-to-car communication, München: BMW group research and technology, http://www.car-to-car.org
2. A. Aijaz, B. Bochow, F. Dötzer, A. Festag, M. Gerlach, R. Kroh, and T. Leinmüller, Attacks on inter vehicle communication systems – an analysis. In Proceedings of Third International Workshop on Intelligent Transportation, WIT 2006, Hamburg, Germany, March 2006
3. M. Gerlach, security challenges for C2C CC standards and relevant research. Presentation at car2car forum, May 2007
4. J.P. Hubaux, S. Capkun, and J. Luo, The security and privacy of smart vehicles. IEEE Security and Privacy, Vol. 4, No. 3, pp. 49–55, 2004
5. B. Parno and A. Perring, Challenges in securing vehicular networks. In Proceeding of Workshop on Hot Topics in Networks (HotNets-IV), November 2005
6. E. Fonseca and A. Festag, A survey of existing approaches for secure ad hoc routing and their applicability to VANETS, NEC network laboratories, March 2006
7. IEEE, IEEE standard 1471–2000: IEEE recommended practice for architectural description of software-intensive systems, 2000
8. http://www.network-on-wheels.de
9. M. Gerlach, NoW – Network on Wheels: C2X security. In Proceedings of the 5th Workshop on Intelligent Transportation Systems, WIT, Hamburg, February 2008
10. M. Gerlach, A. Festag, T. Leinmüller, G. Goldacker, and C. Harsch, Security architecture for vehicular communication, NoW project, WIT 2007
11. http://www.sevecom.org
12. A. Kung, Security architecture and mechanisms for V2V/V2I. Public deliverable D2.1, Sevecom project, 2007
13. http://www.gstforum.org
14. http://evita-project.org
15. http://www.preciosa-project.org

# Index

Note: Page numbers followed by 'f' and 't' indicate figures and tables respectively.




R. Popescu-Zeletin et al., *Vehicular-2-X Communication*,
DOI 10.1007/978-3-540-77143-2, © Springer-Verlag Berlin Heidelberg 2010

 MIX
Papier aus verantwortungsvollen Quellen
Paper from responsible sources
FSC® C105338

If you have any concerns about our products,
you can contact us on
ProductSafety@springernature.com

In case Publisher is established outside the EU,
the EU authorized representative is:
**Springer Nature Customer Service Center GmbH
Europaplatz 3, 69115 Heidelberg, Germany**

Printed by Libri Plureos GmbH
in Hamburg, Germany